Advanced Biomass Gasification

Advanced Biomass Gasification

New Concepts for Efficiency Increase
and Product Flexibility

Steffen Heidenreich
Michael Müller
Pier Ugo Foscolo

AMSTERDAM • BOSTON • HEIDELBERG • LONDON
NEW YORK • OXFORD • PARIS • SAN DIEGO
SAN FRANCISCO • SINGAPORE • SYDNEY • TOKYO

Academic Press is an imprint of Elsevier

British Library Cataloguing-in-Publication Data
A catalogue record for this book is available from the British Library

Library of Congress Cataloging-in-Publication Data
A catalog record for this book is available from the Library of Congress

ISBN: 978-0-12-804296-0

For Information on all Academic Press publications
visit our website at http://www.elsevier.com/

Working together
to grow libraries in
developing countries

www.elsevier.com • www.bookaid.org

Publisher: Joe Hayton
Acquisition Editor: Raquel Zanol
Editorial Project Manager: Ana Claudia A. Garcia
Production Project Manager: Sruthi Satheesh
Designer: Vicky Pearson Esser

Typeset by MPS Limited, Chennai, India

Contents

Chapter 1

Introduction

Considerate exploitation of the available natural resources is a key issue in the sustainable supply of energy in terms of heat, power, and fuels. In this context, the utilization of renewable energy sources is a major contribution. Moreover, global warming and climate change concerns are resulting in efforts to reduce CO_2 greenhouse gas emissions by increasing the use of renewable energies and increasing the energy efficiency. Besides solar, wind, and hydro energy, biomass is considered as the main renewable energy source. As a renewable solid fuel it is suitable to replace fossil solid fuels like hard coal and lignite. In a renewable energy mixture with fluctuating availability of solar and wind energy, biomass can be exploited as a storable and adjustable energy source that will be used in increased amounts when wind and solar energy supply is low. Therefore, several developed as well as developing countries all over the world have set targets for the share of biomass to the national energy supply and have introduced policies to promote the increasing use of biomass as an energy source.

Since the discovery by mankind of how to make fire, biomass has been the main energy source for thousands of years and still today it contributes in the range of more than 10% to the world energy supply and ranks as the fourth source of energy in the world [1]. In rural agricultural areas, biomass is still the main energy resource for heating and cooking and often it is the only available energy source. In developing countries in Asia and Africa more than one-third of the total energy consumption is based on biomass. A big advantage of biomass is its availability at every place all over the world which is in contrast to fossil fuels like coal, oil, or natural gas. By way of example, India has very large coal reserves of more than 250 billion tons in the state of Bihar and northeast. However, transportation costs play a major role in the distribution of the coal all over the country. In contrast, biomass is uniformly and widely distributed over the whole country [2].

Beside combustion of biomass for production of heat and power, which is still the main energetic utilization of biomass, gasification is a key technology for the use of biomass. It offers the advantage of a high flexibility in using different kinds of feedstock materials as well as in the generation of different products. In principal, all different types of biomass can be converted by gasification into a product gas mainly consisting of hydrogen, carbon monoxide, carbon

Advanced Biomass Gasification. DOI: http://dx.doi.org/10.1016/B978-0-12-804296-0.00001-4

dioxide, and methane. From this product gas, all kinds of energy or energy carriers, for example, heat, power, biofuels, hydrogen, and biomethane, as well as chemicals, can be provided. Synthesis of Fischer-Tropsch diesel, dimethyl ether, methanol, and methane from synthesis gas are established technical processes. The use of the available biomass resources needs to be highly efficient and sustainable. Gasification offers high potential and high process efficiency for the use of biomass [3].

Gasification of biomass is performed by partial oxidation of the carbon contained in the biomass at high temperature using a controlled amount of an oxidant which can be either air, pure oxygen and steam, or a mixture of several gasification agents. The yield and composition of the product gas depend on the biomass feedstock, the gasifier type, and the operation conditions of the gasifier, such as the used gasification agent, the temperature, and the residence time in the gasifier.

Biomass comprises a broad range of different kinds of bio materials, such as wood, forest and agricultural residues, waste from wood and food industry, algae, energy grasses, straw, bagasse, sewage sludge, etc. The use of different kinds of biomass results in different challenges and solutions for transportation, storage, pretreatment and feeding of the biomass, for operation of the gasifier, and for cleaning of the produced syngas. Most commonly used types of biomass gasifiers are fixed bed and moving bed, fluidized bed, and entrained flow gasifiers.

Depending on the use of the syngas, its cleaning needs to be very efficient. Catalytic synthesis reactions or its use in fuel cells, for example, require high purity of the syngas. The main impurities in the syngas are fly ash particles and tar. Other impurities in the syngas are typically sulfur compounds (eg, H_2S, COS), hydrogen chloride, alkali compounds, and ammonia. Tar formation is a main problem in biomass gasification. Tar condensation at lower temperatures can cause clogging or blockage of pipes, filters, catalyst units, or engines. Tar formation also lowers the syngas yield and the heating value of the gas. Tar removal has been the subject of much research leading to the development of primary and secondary measures for tar reduction. Overviews on this topic have been recently given, for example, see Han and Kim [4], Aravind and de Jong [5], and Shen and Yoshikawa [6].

In order to promote the utilization of biomass gasification, advanced concepts are required which have to maximize the syngas yield, optimize the gas quality, increase the gas purity, increase the overall process efficiency, and improve the economic viability by decreasing system and production costs.

This book aims at providing an overview on such new concepts in biomass gasification. After a short introduction to fundamental concepts and pretreatment options, concepts for process integration and combination, new and improved gasification concepts, as well as polygeneration strategies are presented.

REFERENCES

[1] Saidur R, Abdelaziz EA, Demirbas A, Hossain MS, Mekhilef S. A review on biomass as a fuel for boilers. Renew Sustain Energy Rev 2011;15:2262–89.

[2] Buragohain B, Mahanta P, Moholkar VS. Biomass gasification for decentralized power generation: the Indian perspective. Renew Sustain Energy Rev 2010;14:73–92.

[3] Ahrenfeldt J, Thomsen TP, Henriksen U, Clausen LR. Biomass gasification cogeneration—a review of state of the art technology and near future perspectives. Appl Therm Eng 2013;50: 1407–17.

[4] Han J, Kim H. The reduction and control technology of tar during biomass gasification/pyrolysis: an overview. Renew Sustain Energy Rev 2008;12:397–416.

[5] Aravind PV, de Jong W. Evaluation of high temperature gas cleaning options for biomass gasification product gas for solid oxide fuel cells. Prog Energy Combust Sci 2012;38:737–64.

[6] Shen Y, Yoshikawa K. Recent progress in catalytic tar elimination during biomass gasification or pyrolysis e a review. Renew Sustain Energy Rev 2013;21:371–92.

Chapter 2

Fundamental Concepts in Biomass Gasification

2.1 CHEMISTRY OF GASIFICATION

Gasification is a thermochemical conversion of a solid or liquid fuel into combustible gases by understoichiometric addition of a gasification agent (oxygen/air, steam, carbon dioxide) at high temperature. The so-called "producer gas" (also called product gas, synthesis gas, or syngas) can be used for heat production, (combined heat and) power generation, and the production of chemicals and fuels [1–3]. Fig. 2.1 shows a general scheme for possible process chains.

The gasification process itself can be divided into several steps, which are heating up of the fuel, drying of the fuel, pyrolysis, and gasification. As a fuel particle is heated, the evaporation of the water contained in the fuel occurs at temperatures above 100°C depending on operation pressure. During devolatilization or pyrolysis, which occurs between 300°C and 600°C, the main organic constituents of the fuel are thermally decomposed into permanent gases, condensable vapors, liquids, and coke. The pyrolysis reactions can be summarized as follows:

$$\text{Fuel} + \text{Heat} \rightarrow \text{Gas } (CH_4, C_mH_n, CO_2, CO, H_2, H_2O, \text{etc.}) + \text{Tars} + \text{Char} \quad (2.1)$$

In the gasification step, the pyrolysis products react further at relatively high temperatures between 700°C and 1500°C with the gasification agent or product gases by numerous chemical reactions. The most important homogeneous gas phase reactions and heterogeneous reactions between solid matter and gases are as follows [4,5]:

Combustion reactions

$$C(s) + O_2 \leftrightarrow CO_2 \quad {}^{''}_r H^0_{298} = -394 \text{ kJ/mol} \quad (2.2)$$

$$C(s) + \tfrac{1}{2}O_2 \leftrightarrow CO \quad {}^{''}_r H^0_{298} = -111 \text{ kJ/mol} \quad (2.3)$$

$$CO + \tfrac{1}{2}O_2 \leftrightarrow CO_2 \quad {}^{''}_r H^0_{298} = -283 \text{ kJ/mol} \quad (2.4)$$

$$H_2 + \tfrac{1}{2}O_2 \leftrightarrow H_2O \quad {}^{''}_r H^0_{298} = -242 \text{ kJ/mol} \quad (2.5)$$

Advanced Biomass Gasification. DOI: http://dx.doi.org/10.1016/B978-0-12-804296-0.00002-6

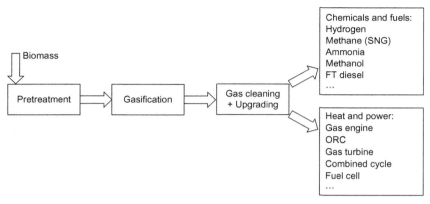

FIGURE 2.1   Pathways for the conversion of biomass to several products. *SNG*, synthetic natural gas; *FT*, Fischer–Tropsch; *ORC*, organic Rankine cycle.

Homogeneous gasification reactions

Water gas shift reaction

$$CO + H_2O \leftrightarrow CO_2 + H_2 \quad \text{''}_r H^0_{298} = -41 \text{ kJ/mol} \tag{2.6}$$

Steam reforming

$$CH_4 + H_2O \leftrightarrow CO + 3H_2 \quad \text{''}_r H^0_{298} = +206 \text{ kJ/mol} \tag{2.7}$$

$$C_n H_m + n H_2O \leftrightarrow nCO + (n + m/2)H_2 \tag{2.8}$$

Dry (CO_2) reforming

$$CH_4 + CO_2 \leftrightarrow 2CO + 2H_2 \quad \text{''}_r H^0_{298} = +247 \text{ kJ/mol} \tag{2.9}$$

$$C_n H_m + n CO_2 \leftrightarrow 2nCO + m/2 H_2 \tag{2.10}$$

Heterogeneous gasification reactions

Boudouard reaction

$$C(s) + CO_2 \leftrightarrow 2CO \quad \text{''}_r H^0_{298} = +172 \text{ kJ/mol} \tag{2.11}$$

Water gas reaction

$$C(s) + H_2O \leftrightarrow CO + H_2 \quad \text{''}_r H^0_{298} = +131 \text{ kJ/mol} \tag{2.12}$$

Methanation

$$C(s) + 2H_2 \leftrightarrow CH_4 \quad \text{''}_r H^0_{298} = -75 \text{ kJ/mol} \tag{2.13}$$

TABLE 2.1 Product Gas Composition and Heating Value for Air- and Oxygen/Steam-Blown Gasification [1,2,6,7]

Gas (vol. %, dry)	Air	Steam/oxygen
H_2	6–22	26–55
CO	9–21	20–47
CO_2	11–19	9–30
CH_4	1–7	4–25
N_2	40–60	0–3
LHV (MJ/m_N^3)	3–6.5	12–17

The product gas mainly consists of the combustible gases hydrogen (H_2), carbon monoxide (CO), and methane (CH_4) and the incombustible gases carbon dioxide (CO_2), water vapor (H_2O), and nitrogen (N_2). Furthermore, a number of undesired trace compounds are present in the product gas, which will be addressed in Section 4.2. The composition of the product gas depends on the type and composition of the biomass, type and amount of gasification agent, type of gasification reactor and residence time, gasification temperature and pressure, presence of catalysts, and several other factors. Table 2.1 gives an overview on the range of product gas compositions and their heating values.

For the further use of the product gas in synthesis of fuels or chemicals, product and process specific ratios of CO to H_2 are desired. The theoretically achievable ratios can be determined by simple thermodynamic calculations [8]. Since biomass feedstocks already contain C, O, and H in a certain ratio, not all ratios of CO to H_2 can be achieved by gasification with oxygen and/or steam as illustrated in Fig. 2.2.

Thermodynamic equilibrium is usually not achieved at gasification temperatures below 1000°C, since residence time and/or mixing are not sufficient in real gasifiers. Thus, producer gases contain higher concentrations of hydrocarbons like methane and tars than predicted by equilibrium calculations. Also, the concentration of ammonia (NH_3) can be higher. Therefore, the actual composition and yield of the producer gas from a specific gasifier can only accurately be predicted by a suitable process model considering the relevant kinetics. These need to include reaction kinetics as well as mass transfer aspects. A detailed description of kinetic and mass transfer aspects in biomass gasification can be found in [10].

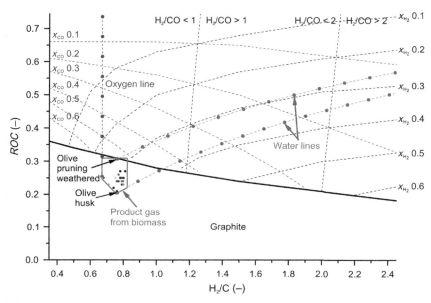

FIGURE 2.2 Achievable producer gas compositions resulting from the gasification of biomass with oxygen and steam at 850°C and 0.1 MPa [9].

2.2 GASIFICATION TECHNOLOGY

A number of different gasification reactors is commercially available or under development. They can be classified in several ways [1–3,11,12]:

According to the heat supply: autothermal or allothermal gasification.
According to the gasification agent: air, oxygen, and/or steam.
According to the transport process within the reactor: fixed bed, fluidized bed, or entrained flow.
According to the pressure in the gasifier: atmospheric or pressurized.

In autothermal or direct gasification, the necessary heat for the endothermic gasification reactions is supplied by partial oxidation of the fuel. In allothermal or indirect gasification, the heat is supplied from an external source, for example, combustion of a part of the gasification products. Due to the partial oxidation, product gases from autothermal gasification have lower heating values than product gases from allothermal gasification.

An overview of the basic principles of different reactors according to the transport processes is shown in Fig. 2.3.

Fixed bed gasifiers are used for small-scale operation of up to several MW$_{th}$. The gasification takes place in a fixed bed, in which the different gasification reactions occur in different zones. The gasifiers are divided into downdraft or

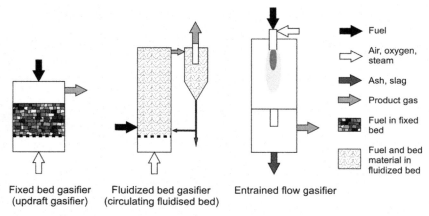

Fixed bed gasifier (updraft gasifier) Fluidized bed gasifier (circulating fluidised bed) Entrained flow gasifier

FIGURE 2.3 Basic principles of gasification processes.

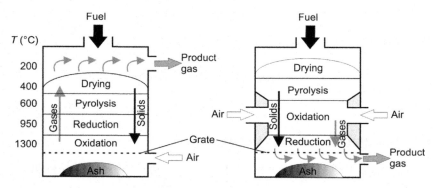

FIGURE 2.4 Updraft gasifier (left) and downdraft gasifier (right).

cocurrent gasifiers and updraft or countercurrent gasifiers. While in both config-urations the fuel is usually fed from the top of the gasifier, the gasification agent, which is usually air, is added at the bottom in case of updraft gasification and in the middle in case of downdraft gasification. As a consequence, a different order of the zones forms in the gasifiers as illustrated in Fig. 2.4. Major advan-tage of updraft gasifiers is their low requirement on fuel quality and size. Major drawback is the high content of tars ($10–150 \, \mathrm{g/m_N^3}$) and particles ($0.1–3 \, \mathrm{g/m_N^3}$) in the producer gas, which is caused by the opposite flow direction of fuel and gas. Thereby, tars produced during pyrolysis will not be cracked or oxidized. Furthermore, the high particle load of the product gas occurs since most of the particles are released in the drying zone. A reduction in content of tars can be achieved by cocurrent flow of solids and gases, as it is the case in downdraft gasifiers. The pyrolysis gases have to pass the oxidation zone, so that tars are thermally cracked or oxidized in the oxidation zone. Thus, the tar content is reduced to $0.1–0.6 \, \mathrm{g/m_N^3}$.

The main design types of fluidized bed gasifiers are bubbling (BFB) and circulating (CFB) fluidized beds, which can have a capacity of more than $100\,MW_{th}$. The fuel is mixed to the bed material and fluidized by the gasification agent. Due to the high gas velocity and the absence of a distinct oxidation zone, the product gas contains relatively high amounts of particles $(1-100\,g/m_N^3)$ and tars $(1-30\,g/m_N^3)$. While in the case of BFB the entrained particles are removed from the process, they are recirculated into the bed in the case of CFB. Therefore, the conversion is higher in CFB. Usually silica sand or catalytically active olivine sand is used as a bed material. Operation temperatures are typically in the range of 800°C. The upper operation temperature needs to be limited to prevent bed agglomeration and thus depends mainly on the ash composition.

Two fluidized beds can be interconnected to achieve a specific gas quality. One commercial example is the fast internal circulating fluidized bed (FICFB) reactor [13] which is described in detail in Section 5.2.1. In the case of FICFB, one fluidized bed is operated as allothermal steam gasification and the other one as combustion to provide the necessary heat for the endothermic gasification reactions. The heat is transported between the two reactors by the bed material.

In entrained flow gasification tar and particle/ash related problems are avoided by very high operation temperatures of up to 1600°C. Due to the high temperatures, tars are effectively cracked and particles are molten, so that they can easily be removed as liquid slag. Furthermore, gasifier slags have a high potential for alkali and heavy metal retention depending on their composition [14,15]. Entrained flow gasifiers are designed for large-scale operation at $>>100\,MW_{th}$.

REFERENCES

[1] Knoef H. Handbook biomass gasification, 2nd ed. Enschede (The Netherlands): Biomass Technology Group; 2012.

[2] Kaltschmitt M, Hartmann H, Hofbauer H. Energie aus Biomasse, 2nd ed. Heidelberg (Germany): Springer; 2009.

[3] Higman C, van der Burgt M. Gasification, 2nd ed. Burlington: Gulf Professional Publishing; 2008.

[4] Jüntgen H, van Heek KH. Kohlevergasung—Grundlagen und Technische Anwendung. München: Verlag Karl Thiemig; 1981.

[5] Higman C, van der Burgt M. Gasification. Amsterdam: Elsevier, Gulf Professional Publishing; 2008.

[6] Gil J, Corella J, Aznar MP, Caballero MA. Biomass gasification in atmospheric and bubbling fluidized bed: effect of the type of gasifying agent on the product distribution. Biomass Bioenerg 1999;16:1–15.

[7] Hofbauer H, Fleck T, Veronik G. Gasification feedstock database. IEA Bioenergy Agreement, Task XIII, Activity 3, Vienna, Austria: Technische Universität Wien; 1997.

[8] Stemmler M, Müller M. Theoretical evaluation of feedstock gasification using H_2/C ratio and ROC as main input variables. Ind Eng Chem Res 2010;49:9230–7.

[9] Stemmler M, Tamburro A, Müller M. Thermodynamic modelling of fate and removal of alkali species and sour gases from biomass gasification for production of biofuels. Biomass Conv Bioref 2013;3:187–98.

[10] De Jong W, van Ommen JR. Biomass as a sustainable energy source for the future: fundamentals of conversion processes. Hoboken (New Jersey): John Wiley & Sons; 2015.

[11] Hofbauer H. Conversion technologies: gasification overview. In: Proceedings of the 15th European Biomass Conference & Exhibition. Berlin, Germany; 2007.

[12] Brown D, Gassner M, Fuchino T, Marechal F. Thermo-economic analysis for the optimal conceptual design of biomass gasification energy conversion systems. Appl Therm Eng 2009;29:2137–52.

[13] Hofbauer H, Rauch R, Loeffler G, Kaiser S, Fercher E, Tremmel H. Six years experience with the FICFB-Gasification process. In: Proceedings of the 12th European Conference on Biomass and Bioenergy. Amsterdam, The Netherlands; 2002. p. 982–5.

[14] Yun YS, Ju JS. Operation performance of a pilot-scale gasification/melting process for liquid and slurry-type wastes. Korean J Chem Eng 2003;20:1037–44.

[15] Willenborg W, Müller M, Hilpert K. Alkali removal at about 1400°C for the pressurized pulverized coal combustion combined cycle. 1. Thermodynamics and concept. Energ Fuel 2006;20:2593–8.

Chapter 3

Biomass Pretreatment

3.1 INTRODUCTION

Biomass is by definition all material of organic origin. Thus, it includes plants and animals, their waste and residues (eg, excreta), and all materials derived by their conversion or utilization (eg, pulp and paper, municipal waste, sewage sludge, vegetable oil, alcohol) [1,2]. Owing to the wide variety of biomass sources, there are a wide variety of chemical and physical properties of biomass. However, conversion processes like gasification usually require specific chemical and physical properties of a fuel. Especially particle size, water content, ash amount and behavior, and reactivity can be crucial properties influencing or even determining the choice of a suitable conversion process and suitable process parameters. Furthermore, stability and energy density are important properties for transport and storage of the biomass. Therefore, fuel properties are very often adjusted by means of physical pretreatment of the biomass feedstock. Depending on the specific process requirements, also chemical or thermochemical pretreatment might be advantageous.

3.2 PHYSICAL PRETREATMENT

Physical pretreatment is an important step, or several steps, in the supply chain of biomass fuels, comprising harvesting, transport, storage, washing, size reduction, drying, and compacting. The steps and order of steps in the supply chain depend on the type and properties of the biomass feedstock as well as the fuel requirements of the respective conversion process. Information on supply chains for woody and herbaceous biomass for utilization as fuel in thermo chemical conversion processes can be found in [1,3]. In the following, important physical pretreatment methods are briefly discussed.

3.2.1 Washing

Washing of the biomass can be used to lower the mineral matter content, especially with respect to inorganic constituents, which can cause problems during thermo chemical conversion like slagging, fouling, corrosion, and bed agglomeration (see Section 4.2.2). In particular, herbaceous biomass contains relatively

Advanced Biomass Gasification. DOI: http://dx.doi.org/10.1016/B978-0-12-804296-0.00003-8

high amounts of water soluble alkali, chlorine, and sulfur compounds. These can be efficiently reduced by leaching with water, resulting in improved ash behavior and reduction of the related problems listed before [4–7]. The easiest way of washing is exposing the harvested herbaceous biomass to rain, so-called weathering, which already shows significant effects [8]. However, with varying weather also the quality of the weathered biomass varies significantly.

3.2.2 Drying

Drying of the biomass can have several advantages, for example, preservation, weight reduction, and increase in heating value. It can be either required by steps of the supply chain, for example, storage and compacting, or the thermochemical conversion process. The water content can be reduced by several technologies, that is, natural drying, mechanical drying, and thermal drying.

The amount of water in biomass varies depending on type, growing conditions, and time of harvesting. The water can be bound in the biomass in different ways with increasing binding forces from capillary sorption and adhesion to chemisorption [9]. These binding forces have to be overcome during drying. Consequently, Colin and Gazbar found water in sludge with different behavior during mechanical dewatering, which they categorized as free water, water removable by moderate mechanical strain, water removable by maximum mechanical strain, and water not removable mechanically [10].

The easiest way to reduce the moisture content of fresh biomass is natural drying by air, where the achievable moisture content of the biomass depends on the temperature and relative humidity of the air. Herbaceous biomass is often just left on the ground to be dried by the surrounding air. The drying process can be accelerated by raking. Thus, the moisture content of straw can be reduced from 40% to below 20% within a few days [11]. Also wood, containing up to 55% moisture after felling [9], can principally be dried that way down to 20% moisture [12]. However, wood is usually stapled for drying to enhance the process by natural convection [13]. Furthermore, the process can be accelerated by splitting of the stems. In bulk goods like wood chips, the natural convection is further enhanced by slight self-warming of the biomass due to digestion of organic matter [14].

Compressible biomass, for example, sludge, can be dried by mechanical dewatering. Mechanical presses include screw, belt filter, ring, drum, and roller presses [15]. Also centrifuges can be used. The major drawback of mechanical drying processes is their energy consumption and maintenance costs.

Even more energy-intensive than mechanical drying is thermal drying. However, if the drying process is located close to the conversion process, waste heat can be used for drying. A number of processes have been developed for industrial application [16]. Wet biomass can either be directly dried by hot air, steam, or flue gas in belt conveyers, fluidized beds, spray dryers, etc., or indirectly by heat-transfer via a casing, for example, in a drum. The selection of the drying process depends on the properties of the biomass, for example, size, morphology, and heat sensitivity, as well as the requirements of the entire process chain.

3.2.3 Size Reduction and Compaction

Raw biomass like herbaceous and woody biomass has usually an unfavorable and wide size distribution (eg, stalks, stems), a low bulk density, and a resulting low volumetric energy density. Therefore, the aim of mechanical treatment like size reduction and compaction of raw biomass is adjusting the feedstock to the requirements of the conversion process regarding size, homogeneity, and physical properties of the fuel.

Since biomass is usually very heterogeneous and often fibrous, size reduction is complex compared to coal or minerals. Shearing, ripping, and cutting actions are needed for size reduction, because biomass often only deforms in crushing processes. Depending on the physical properties of the raw biomass, techniques like chipping, shredding, crushing, and milling are used. Kaltschmitt [1] gives a good overview on the several methods used for size reduction. By way of example, hammer mills show good performance in milling of herbaceous biomass [17].

Major methods for compacting are briquetting and pelletizing (Fig. 3.1). Both methods require milling and drying of the biomass before further processing. In case of briquetting, the milled raw material is compacted under high pressure without binder material using continuously working screw or piston presses. Due to friction, the temperature increases causing softening of the lignin which thus can act as binder. In case of pelletizing, the milled raw material is typically pressed (with binder) by a kollergang press through a flat or ring matrix. The produced pellets have a diameter of up to 25 mm, typically 5–12 mm, and a length of 3–50 mm [18].

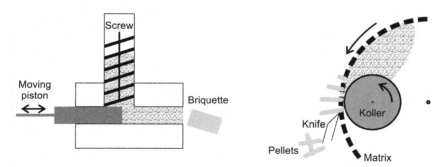

FIGURE 3.1 Principle of a piston press (left) for briquetting and a kollergang press for pelletizing.

3.3 TORREFACTION

Torrefaction is a thermal treatment of biomass at relatively low temperatures between 200°C and 300°C under atmospheric pressure in the absence of oxygen [19–21]. The process is also called roasting, cooking, high temperature drying, and mild pyrolysis. The aim of the process is the improvement of physical and chemical properties of the fuel. Torrefaction has been investigated with a broad variety of biomass feedstock covering woody as well as herbaceous biomass.

Bergman et al. defined five process steps characterizing the entire torrefaction process. During initial heating, the biomass is heated up to the temperature moisture starts to evaporate. During predrying at 100°C, the free water evaporates. During postdrying and intermediate heating at temperatures up to 200°C, physically bound water and volatile organic compounds evaporate and the formation of CO_2 starts. Torrefaction reactions occur at temperatures above 200°C, while the torrefaction temperature is defined as the maximum constant temperature. The last step is cooling of the solids.

The torrefaction reactions comprise reactions of the three main biomass constituents, namely hemicellulose, cellulose, and lignin [22]. Depending on molecular structure, hemicellulose starts to decompose into volatiles and char at temperatures above 200°C, leading to extensive devolatilization at temperatures above 250°C. Already below 200°C, cellulose starts to depolymerize, which is believed to be the main reason that biomass loses its tenacity and structure [20]. Above 270°C, noticeable reactions of cellulose and lignin occur [22]. However, hemicellulose, cellulose, and lignin react independently and do not influence each other [23].

The main effect of torrefaction on the chemical composition of biomass is the reduction in oxygen content. By losing oxygen and also hydrogen, torrefied biomass becomes more like coal, as can be seen in the van Krevelen diagram in Fig. 3.2, and

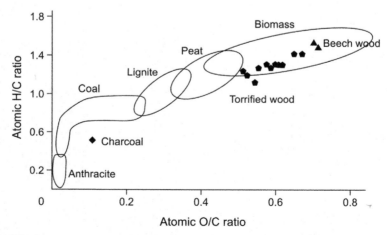

FIGURE 3.2 Van Krevelen diagram. Data for wood-based fuels are taken from [25].

the calorific value (lower heating value (LHV)) of the fuel increases [24,25]. In the case of wood, the oxygen content decreases from 45.1% to 36.3% by torrefaction at 300°C, while the LHV is increased from 17.6 to 21 MJ/kg. Depending on temperature, more than 70% of the mass is retained in the solid product. Furthermore, by destruction of the fibrous structure and tenacity of biomass during torrefaction, the fuel becomes more brittle [26], resulting in an improved grindability of the fuel [27]. In addition, the fuel becomes more hydrophobic [26], resulting in significantly improved stability against rotting.

The torrefaction process can be performed in different types of reactors, for example, rotating drum, screw conveyer, or moving bed, depending on the type, quality, and amount of the biomass. A major issue in process design is heat integration. Though the torrefaction reactions are not very exo- or endothermic, the vaporization of the water requires a certain amount of energy.

3.4 HYDROTHERMAL CARBONIZATION

Hydrothermal carbonization (HTC) is a thermochemical treatment of biomass in pressurized water at relatively low temperatures between 180°C and 250°C at or above saturated pressure. The aim of the process is the conversion of biomass into a coal-like fuel, which is therefore often called biocoal or HTC-coal [28]. HTC is a favorable pretreatment method for wet biomass, since the process takes place in water. Therefore, in addition to woody and herbaceous biomass also wet biomass like moss and grass have been treated hydrothermally [29].

Important reactions during hydrothermal treatment are decarboxylation, hydrolysis, and dehydration resulting mainly in a decrease in O/C ratio as shown in the van Krevelen diagram in Fig. 3.3 [30–32]. Carbon losses result more from dissolving soluble organics, for example, sugars, acids, and phenols, into the

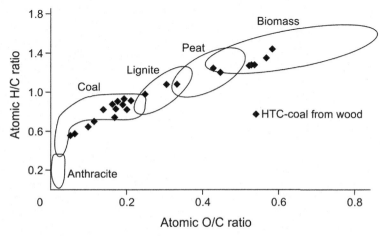

FIGURE 3.3 Van Krevelen diagram. Data for wood-derived HTC-coal are taken from [29].

water than vaporization of volatiles [33,34]. Depending on biomass feedstock, the coal yield is between 35% and 80%. Additionally, reactions of the solids similar to those occurring during torrefaction, for example, depolymerization of cellulose, have to be considered due to similar process temperatures. The product composition depends on the feedstock, the temperature, and the reaction time. The degree of carbonization increases with increasing temperature and reaction time.

From different biomass feedstock HTC-coals with a higher heating value of about 27 MJ/kg, H/C = 1 and O/C = 0.3, which is very close to lignite, were produced after 5–40 h at 200–250°C [29]. At 275°C and 3 h even values comparable to hard coal can be achieved. Furthermore, HTC-coals show good grindability, which is related to the depolymerization of cellulose, and an increased mechanical dewatering capability. During gasification experiments, HTC-coals showed behavior comparable to lignite [35].

REFERENCES

[1] Kaltschmitt M, Hartmann H, Hofbauer H. Energie aus Biomasse, 2nd ed. Heidelberg (Germany): Springer; 2009.

[2] Kleemann M, Meliß M. Regenerative Energiequellen. Berlin (Germany): Springer; 1993.

[3] Kaltschmitt M, Thrän D. Logistik für die Versorgung von Anlagen zur energetischen Nutzung biogener Festbrennstoffe—Anforderungen und Randbedingungen. Zeitschrift für Energiewirtschaft. Wiesbaden: Friedr. Vieweg & Sohn Verlagsgesellschaft mbH; Heft 4/2006.

[4] Jenkins BM, Bakker RR, Wei JB. On the properties of washed straw. Biomass Bioenerg 1996;10:177–200.

[5] Turn SQ, Kinoshita CM, Ishimura DM. Removal of inorganic constituents of biomass feedstock by mechanical dewatering and leaching. Biomass Bioenerg 1997;12:241–52.

[6] Jenkins BM, Baxter LL, Miles TR. Combustion properties of biomass. Fuel Proc Technol 1998;54:17–46.

[7] Arvelakis S, Vourliotis P, Kakaras E, Koukis EG. Effect of leaching on the ash behavior of wheat straw and olive residue during fluidized bed combustion. Biomass Bioenerg 2001;20:459–70.

[8] Van Loo S, Koppejan J, editors. Handbook of biomass combustion and co-firing. Enschede (The Netherlands): Twente University Press; 2004.

[9] Brusche R. Hackschnitzel aus Schwachholz. KTBL Schrift 290. Münster (Germany): Landwirtschaftsverlag; 1983.

[10] Colin F, Gazbar S. Distribution of water in sludges in relation to their mechanical dewatering. Water Res 1995;29:2000–5.

[11] Hartman H, Strehler A. Die Stellung der Biomasse im Vergleich zu anderen erneuerbaren Energieträgern aus ökologischer, ökonomischer und technischer Sicht. Schriftenreihe Nachwachsende Rohstoffe, Band 3. Münster (Germany): Landwirtschaftsverlag; 1995.

[12] Weingartmann H. Hackguttrocknung. Landtechnische Schriften Nr. 178. Wien (Austria): Österreichisches Kuratorium für Landtechnik; 1991.

[13] Höldrich A, Hartmann H, Decker T, Reisinger K, Schardt M, Sommer W, et al. Rationelle Scheitholzbereitstellungsverfahren. Berichte aus dem TFZ, Nr. 11, Technologie- und Förderzentrum (TFZ). Straubing (Germany): Selbstverlag; 2006.

[14] Feller S, Webenau B, Weixler H, Krausenboeck B, Güldner A, Remler N. Teilmechanisierte Bereitstellung, Lagerung und Logistik von Waldhackschnitzeln. LWF-Schriftenreihe Nr. 21. Freising (Germany): Bayerische Landesanstalt für Wald und Forstwirtschaft; 1999.

[15] Roos CJ. Biomass drying and dewatering for clean heat and power. Olympia (WA): Northwest CHP Application Center; 2008.

[16] Mujumdar AS, editor. Handbook of industrial drying. New York: CRC Press; 2006.

[17] Mani S, Tabil LG, Sokhansanj S. Grinding performance and physical properties of wheat and barley straws, corn stover and switchgrass. Biomass Bioenerg 2004;27:339–52.

[18] Deutsches Institut für Normung   CEN TS 14961 (Feste Biobrennstoffe—Brennstoffspezifikationen und -klassen). Berlin (Germany): Beuth; 2005.

[19] Bergman PCA, Boersma AR, Zwart RWR, Kiel JHA. Torrefaction for biomass co-firing in existing coal-fired power stations "biocoal". Report ECN-C-05-013. Petten (The Netherlands): ECN; 2005.

[20] Bergman PCA, Boersma AR, Kiel JHA, Prins MJ, Ptasinski KJ, Janssen FJJG. Torrefaction for entrained flow gasification of biomass. Report ECN-RX-04-046. Petten (The Netherlands): ECN; 2004.

[21] Van der Stelt MJC, Gerhauser H, Kiel JHA, Ptasinski KJ. Biomass upgrading by torrefaction for the production of biofuels: a review. Biomass Bioenerg 2011;35:3748–62.

[22] Chen WH, Kuo PC. A study on torrefaction of various biomass materials and its impact on lignocellulosic structure simulated by a thermogravimetry. Energy 2010;35:2580–6.

[23] Chen WH, Kuo PC. Torrefaction and co-torrefaction characterization of hemicellulose, cellulose and lignin as well as torrefaction of some basic constituents in biomass. Energy 2011;36:803–11.

[24] Prins MJ, Ptasinski KJ, Janssen FJJG. More efficient biomass gasification via torrefaction. Energy 2006;31:3458–70.

[25] Prins MJ, Ptasinski KJ, Janssen FJJG. Torrefaction of wood: part 2. Analysis of products. J Anal Appl Pyrolysis 2006;77:35–40.

[26] Bourgeois J, Guyonnet R. Characterization and analysis of torrefied wood. Wood Sci Technol 1988;22:143–55.

[27] Arias B, Pevida C, Fermoso J, Plaza MG, Rubiera F, Pis JJ. Influence of torrefaction on the grindability and reactivity of woody biomass. Fuel Proc Technol 2008;89:169–75.

[28] Möller M, Nilges P, Harnisch F, Schröder U. Subcritical water as reaction environment: fundamentals of hydrothermal biomass transformation. Chem Sus Chem 2011;4:566–79.

[29] Funke A, Ziegler F. Hydrothermal carbonization of biomass: a literature survey focussing on its technical application and prospects Proceedings of 17th European biomass conference and exhibition. Florence (Italy): ETA-Florence; 2009.

[30] Dinjus E, Kruse A, Tröger N. Hydrothermal carbonization—1. Influence of lignin in lignocelluloses. Chem Eng Technol 2011;34:2037–43.

[31] Falco C, Baccile N, Titirici MM. Morphological and structural differences between glucose, cellulose and lignocellulosic biomass derived hydrothermal carbons. Green Chem 2011;13:3273–81.

[32] Bobleter O. Hydrothermal degradation of polymers derived from plants. Prog Polym Sci 1994;19:797–841.

[33] Berge ND, Ro KS, Mao J, Flora JRV, Chappell MA, Bae S. Hydrothermal carbonization of municipal waste streams. Environ Sci Technol 2011;45:5696–703.

[34] Hoekman SK, Broch A, Robbins C. Hydrothermal carbonization (HTC) of lignocellulosic biomass. Energy Fuels 2011;25:1802–10.

[35] Tremel A, Stemann J, Herrmann M, Erlach B, Spliethoff H. Entrained flow gasification of biocoal from hydrothermal carbonization. Fuel 2012;102:396–403.

Chapter 4

Advanced Process Integration: The *UNIQUE* Gasifier Concept— Integrated Gasification, Gas Cleaning and Conditioning

Decomposition of tar (high molecular weight hydrocarbons) and removal of particles (char, fly ash) from the produced syngas are key issues of gas cleaning in biomass gasification. In existing gasification plants, removal of particulates and tar from the raw syngas is normally performed by filtration and scrubbing. In this way the cleaned produced gas is cooled down to temperatures close to ambient, and the most immediate option for power generation is a gas engine. Such a process configuration limits electric conversion efficiencies—reported values are close to 25% [1], this is what can also be obtained with modern combustion plants coupled with steam turbines. This penalizes notably the overall economic balance of the plant, which would benefit from a higher share of electricity against heat production, due also to the incentives for green electricity offered in most countries. In addition, tar separation is sometimes not as effective as it should be, reduces the gas yield, and generates waste streams that are difficult to dispose of or recycle properly.

High temperature gas cleaning and catalytic conditioning is the focal point to promote more efficient industrial applications of biomass gasification for energy and chemicals. Gas treatments should be strictly integrated with biomass conversion and carried out at a close temperature range, to preserve the thermal energy content of the biomass gas. This is even more true in the case of steam gasification and its coupling with a high temperature fuel cell or downstream catalytic processes, to avoid loss by condensation of the significant amount of water vapor contained in the gas stream, which is useful to reform CH_4, shift CO towards H_2, and prevent carbon deposition on the catalytic surfaces.

The application of such fundamental process integration concepts not only allows to realize more simple, efficient, and cost-effective gasification plants, but offers the opportunity to overcome some major obstacles still preventing a large market diffusion of such technologies, specifically at the small to medium

Advanced Biomass Gasification. DOI: http://dx.doi.org/10.1016/B978-0-12-804296-0.00004-X

scales (less than $10\,MW_{th}$) amenable to the vast economic contexts of developed and developing countries. It needs to be stressed that the integration of a biomass steam gasifier with a high temperature fuel cell (more specifically, a solid oxide fuel cell (SOFC)) appears most suitable to realize very efficient bioenergy systems at a relatively small scale. It is almost universally recognized that power generation by means of a fuel cell can compete in terms of efficiency with large integrated gasification combined cycle installations; high temperature fuel cells are able to utilize the major fuel species obtained from the biomass gasification process (H_2, CO, CH_4). Although they are much more resistant to contaminants than low temperature fuel cells, provision should be made however to drastically reduce the fuel gas content of alkali and sour gas compounds.

Recent developments in innovative catalysts, sorbents, and high temperature filtration media offer the opportunity to integrate in one reactor biomass gasification and gas cleaning and conditioning processes needed to obtain a clean fuel gas from biomass that would allow immediate and efficient conversion into power (high temperature fuel cells; micro gas turbines; combined, strictly integrated heat and power plant schemes) and further catalytic gas processing addressed to second-generation biofuels (liquid fuels, hydrogen) and chemicals, allowing to implement diversified polygeneration strategies.

The *UNIQUE* gasification technology, originally outlined in patents [2,3] and intensively investigated in the framework of a joint European R&D project [4] and applied in a subsequent development project [5] with efforts by several research organizations and private companies throughout Europe, provided the opportunity to develop technical innovations addressed to existing and new industrial installations. Their effectiveness was tested at real industrial conditions, over a significant range of scales, from lab facilities to output slip streams of industrial gasifiers and to a first full size pilot installation of a $1\,MW_{th}$ gasifier.

4.1   THE *UNIQUE* GASIFIER CONCEPT

The UNIQUE gasifier concept is based on the integration of produced gas cleaning and conditioning directly into the gasification reactor. The main item of this concept is the integration of catalytic filter elements for particle and tar removal into the freeboard of a fluidized bed steam gasifier. Catalytic filter elements are a very promising technology to combine particle and tar removal at high temperatures (see Section 4.5). Performing the filtration at high temperature in the freeboard of the fluidized bed gasifier is an energy efficient solution, since catalytic tar reforming requires high temperatures of typically above 800°C. In this way reheating of the gas to the reaction temperature and the need for auxiliary heating of pipes and vessels downstream of the gasifier (filter vessel and catalyst reactor) is prevented. Furthermore, plugging of system components by tar condensation as well as catalyst deactivation by deposition of particles on the catalytic active sites is prevented by performing combined particle and tar removal by using catalytic filter elements at high temperatures. By the integration of catalytic filter elements directly into the gasifier, a very compact integrated fluidized bed gasification and hot gas cleaning and conditioning system in one reactor vessel can be realized. One additional option to increase the gas cleaning performance of the system is to add catalytically active materials and sorbents into the fluidized bed for primary tar reforming (see Section 4.4) and removal of detrimental inorganic trace elements (see Section 4.6), respectively. As a result, some major advantages of traditional primary and secondary hot gas treatments are combined together, without their well-known respective drawbacks (catalyst clogging by solid particles, loss of gas chemical and thermal energy, etc.). By performing a complete hot gas cleaning and conditioning, the energy efficiency of the process is high, since no cooling step is included. Fig. 4.1 shows schematically the principle of this new compact gasifier concept.

Remarkable system simplification and process intensification is achieved by housing the whole gas cleaning and conditioning system in the gasifier, allowing for a very compact unit, increasing the energy efficiency by reducing thermal losses and preventing reheating of the gas for secondary tar reforming, saving equipment and plant space. By integration of the hot gas cleaning and conditioning system into the fluidized bed gasifier, the investment costs of the gas cleaning and conditioning equipment and correspondingly the costs of a

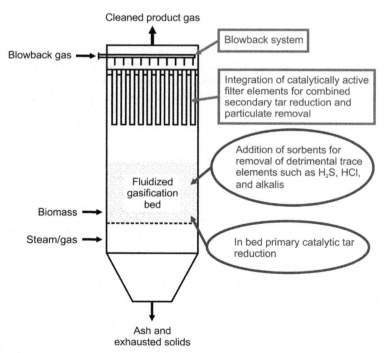

FIGURE 4.1 Scheme of the compact integrated UNIQUE gasifier concept.

biomass gasification plant can be reduced. The arrangement of integration of catalytic filtration and biomass gasification in one reactor vessel offers an efficient reduction of tar, elimination of trace elements, and an efficient abatement of particulates, delivering a high purity syngas, suitable to assure a high share of power generation even in small- to medium-scale (few MW_{th}) combined heat and power production and power plants, and to increase the overall economic revenue of a biomass gasification plant. The UNIQUE gasifier concept for hot gas cleaning and conditioning provides a concrete contribution to the target of reducing the cost of electricity obtained by means of advanced biomass energy systems.

4.2   GAS IMPURITIES AND RELATED PROBLEMS

Beside the major components, CO, H_2, CO_2, H_2O, and CH_4, fuel gases derived from biomass gasification also contain gas impurities like tars, mainly polyaromatic hydrocarbons, and particulate matter (eg, ash, char, soot, bed material). Furthermore, inorganic trace elements contained in the biomass are released during gasification [6–9]. The most important impurities are chlorine compounds (eg, HCl), sulfur species (eg, H_2S, COS, thiophenes, mercaptanes), nitrogen compounds (eg, NH_3, HCN, pyrroles, pyridines), alkali metal species (eg, KCl, KOH, NaCl, NaOH), and other trace metals.

The type and amount of inorganic trace species in the biomass varies with plant type and growing conditions, for example, climate and composition of soil [10]. Furthermore, the time of harvesting impacts the content of trace elements in the plant. Consequently, not only the process conditions determine the release of inorganic trace elements during gasification, but also the composition of the fuel.

Stemmler et al. investigated the influence of biomass composition on the release of alkali species and sour gases during steam gasification at 850°C for a number of biomass fuels by thermodynamic equilibrium calculations [11]. Although equilibrium calculations cannot consider the nonequilibrium state in fluidized bed gasification, the calculated producer gas composition was in relatively good agreement with the measured composition from wood gasification. Furthermore, thermodynamic equilibrium calculations have been shown to be suitable for the simulation of biomass producer gas before [12,13]. Thus, the predicted concentration of impurities gives at least a good indication on the influence of fuel composition on the achievable gas quality. The composition of the biomass fuels [14] is given in Table 4.1 and the predicted concentration of alkali species and sour gases in the producer gas is given in Fig. 4.2.

The different release behavior reflects the different composition of the biomasses. The calculated KCl concentration in the producer gas varies between 25 ppmv for pine seed shells (PiSeSh) and 900 ppmv for corn stover (CoSt). Herbaceous biomasses like CoSt have by far the highest amount of alkalis, chlorine, and sulfur. Therefore, they show by far the highest release of corresponding species, except for KOH which is the highest for woody biomass. However,

TABLE 4.1 Composition of the Biomasses Used for Calculations (wt%)

Biomass	C	H	N	O	Cl	S	Si	Al	Mg	Ca	Na	K	P	Fe
Almond shells UnivAq	48.9	6.2	0.18	43.5	0.029	0.026	0.106	0.014	0.046	0.226	0.010	0.524	0.032	0.013
Almond shells ENEA	47.8	6.38	0.44	43.6	0.018	0.028	0.109	0.020	0.039	0.178	0.017	0.578	0.044	0.017
Pine seed shells	51.3	6.4	0.48	40.7	0.019	0.013	0.160	0.002	0.030	0.028	0.004	0.062	0.006	0.004
Hazel nut shells	51.0	6.1	0.45	41.5	0.026	0.022	0.028	0.006	0.027	0.175	0.001	0.248	0.009	0.006
Wood pellets (Güssing)	49.6	6.3	0.32	43.2	<0.01	0.024	0.012	0.002	0.007	0.078	0.001	0.018	0.003	0.002
Wood chips (Güssing)	48.5	6.2	0.45	43.9	0.014	0.032	0.016	0.002	0.028	0.259	0.001	0.065	0.013	0.002
Wood chips (TUV)	48.4	6.4	0.41	44.2	<0.01	0.043	0.016	0.002	0.025	0.168	0.001	0.069	0.012	0.003
Willow wood chips (Güssing)	47.9	6.3	0.44	44.5	0.017	0.043	0.020	0.003	0.032	0.365	0.001	0.096	0.019	0.003
Pellets (UPT)	49.0	6.52	0.14	42.6	<0.01	0.022	0.127	0.062	0.036	0.163	0.016	0.098	0.008	0.037
Poplar	48.5	6.30	0.25	42.8	0.04	0.03	0.104	0.015	0.033	0.286	0.004	0.096	0.017	0.013
Turkey oak	49.3	6.41	0.14	43.2	0.015	0.08	0.016	0.005	0.023	0.684	0.001	0.095	0.011	0.005
Oak	47.4	6.4	0.51	43.9	0.012	0.043	0.036	0.011	0.085	0.176	0.004	0.125	0.023	0.010
Olive pruning (fresh)	42.2	5.4	0.5	47.7	<0.01	0.012	0.058	0.015	0.050	0.339	0.097	0.104	0.054	0.008
Olive pruning (weathered)	49.1	6.52	0.29	42.4	<0.01	0.014	0.241	0.022	0.077	0.459	0.112	0.123	0.068	0.019
Olive husk	50.9	6.51	0.32	38.4	0.02	0.024	0.043	0.003	0.015	0.100	0.001	0.165	0.023	0.037
Core Kenaf	48.0	6.37	0.35	42.1	0.06	0.07	0.143	0.028	0.155	0.402	0.006	0.338	0.078	0.016
Switchgrass	45.9	6.21	0.67	41.2	0.05	0.08	0.849	0.060	0.289	0.551	0.032	0.887	0.087	0.035
Rice husk	42.8	5.76	0.46	36.2	0.03	0.04	4.728	0.005	0.037	0.074	0.003	0.220	0.037	0.009
Wheat straw	46.1	5.28	0.34	41.7	0.06	0.09	2.298	0.015	0.068	0.302	0.054	0.666	0.079	0.011
Corn stover	44.5	5.01	0.5	40.2	0.44	0.23	1.913	0.424	0.269	0.294	0.023	1.254	0.112	0.223

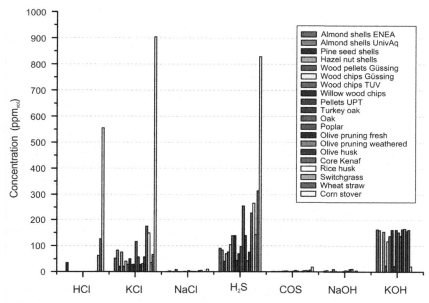

FIGURE 4.2 Calculated concentration of alkali, sulfur, and chlorine compounds in the producer gas.

the release behavior cannot only be explained by the respective concentration of trace elements in the feedstock. In fact, the ratio of different components has to be taken into consideration. If the alkali content exceeds the "free" capacity of aluminosilicates (aluminosilicates preferably react with alkaline earth oxides), alkali compounds will be released. This is the case for all investigated fuels. If no or too little chlorine is available in the biomass, the alkalis will be released as hydroxides, which is the case for the investigated woody biomasses. If the chlorine content is sufficient or even exceeds the available alkali content, the alkalis will be released as chlorides. In the case of chlorine excess, chlorine will be released as HCl in addition. This is the case for the investigated herbaceous biomasses. These findings are in very good agreement with results obtained from release measurements [9].

Due to the reducing conditions, sulfur is released as H_2S and COS. The resulting H_2S concentration in the producer gas is calculated to be 40 ppmv for OlPrWe and 800 ppmv for CoSt. The amount of released sulfur species does mainly correlate with the amount of sulfur in the biomasses, indicating that none of the biomasses has a high potential for sulfur retention. For most biomasses, the resulting H_2S partial pressure is too low to form sulfides with calcium. In the case of the herbaceous biomasses with high sulfur content, calcium is mainly bound in silicates and thus not available for sulfide formation.

The released inorganic trace species like alkali compounds and sour gases can cause or even accelerate problems like high temperature corrosion, fouling, bed agglomeration, and deactivation of catalysts inside the gasifier and downstream equipment.

Since it can alter the properties of the materials, high temperature corrosion is a major problem affecting all hot material surfaces in the process. High temperature corrosion is not only caused by the gas impurities but also by the main producer gas constituents. Furthermore, temperature and oxygen partial pressure play a significant role in corrosion mechanisms. Beside oxidation, carbonization, chlorination, and sulfidation may occur during biomass gasification. For a comprehensive overview on high temperature corrosion the reader is referred to respective textbooks [15–17]. In the following, only the main mechanisms relevant for gasification of biomass will be briefly introduced.

Carbonization occurs independently of trace elements due to an increased carbon activity in the producer gas depending on the CO_2/CO-ratio. It causes the formation of iron and chromium carbides within the protective oxide layer of the steel, thus decreasing the oxidation resistance of the material. The caused embrittlement of the material results in the worst case in the so-called "metal dusting."

Sulfidation of nickel- and iron-based materials occurs in the temperature range between 300°C and 700°C [18]. Beside nickel sulfide (NiS) and iron sulfide (FeS) also chromium sulfide (Cr_2S_3) forms. Since there is a higher diffusion rate of sulfur within the alloy than in the protective oxide layer, the corrosion rate increases after failure of this protective layer.

There are usually oxidizing conditions present in the gasifier during start up and shut down processes causing the formation of SO_x instead of H_2S and sulfation of deposited alkali chlorides. Therefore, also hot corrosion of nickel-based alloys needs to be considered. Due to the relatively low temperatures of fluidized bed gasification there is mainly a risk of hot corrosion type II, which usually occurs at temperatures below 800°C at high SO_3 partial pressure. Under these conditions, nickel sulfate is thermodynamically stable forming with the alkali sulfates in the deposit a eutectic melt (eg, $T_{eutectic}(Na_2SO_4-NiSO_4)$ = 671°C), thus significantly increasing the corrosion rate.

In the case of chlorination, iron chloride ($FeCl_2$) is formed by complex reactions of steels with several gas constituents like HCl, Cl_2, O_2, and H_2O. The corrosion rate depends on the amount of chlorine as well as the temperature of the material.

Catalysts are often used for upgrading the producer gas to valuable products. Since the producer gas is filtered downstream of the gasifier, deactivation of the catalysts by deposition of particulate matter (fouling) is already minimized. However, fouling of the catalyst surface is still possible by condensation of alkali species, occurring already at temperatures above 800°C [19]. As a consequence, the active surface of the catalyst is decreased resulting in an overall decrease in activity. Besides fouling, chemical reactions of the catalyst with producer gas constituents can cause deactivation of the catalyst. The most important reactions are sulfidation and chlorination of the catalyst material [20–22].

Of course, the presence of H_2S is expected to cause not only serious problems for downstream chemical syntheses, but also at a fuel cell anode, especially for fuel cell operation at temperatures below 1000°C [18]. Furthermore, HCN and NH_3 deactivate catalysts used for synthesis process, for example, Fischer–Tropsch catalysts [23].

Depending on biomass composition and temperature profile of the gasifier, alkali metals promote bed agglomeration and formation of ash deposits besides fouling caused by condensation of alkali compounds. These problems are specifically relevant for the gasification of herbaceous biomass containing not only high amounts of alkali salts but also relatively high amounts of silica, causing a low ash softening temperature and melt formation [24]. Partly or even completely molten ash particles hamper removal by hot gas filter candles, since they block pores and react with the candle material [25,26]. Furthermore, slagging and fouling of heat exchanger surfaces may occur reducing the heat transfer.

4.3 ABATEMENT OF PARTICULATES

Fine particulates entrained by the produced gas stream are generated or remain as residues from the feedstock material during the gasification. Additionally, attrition of bed material can contribute to the particulate load of the syngas. The particulates entrained by the syngas can pollute downstream equipment or can block gas pipes. Particle concentrations in the raw syngas from fluidized bed biomass gasification are typically in the order of about $10\,\mathrm{g/m_N}^3$. Since the concentration limits of most electrical energy generating devices or fuel generating process are between 1 and $5\,\mathrm{mg/m_N}^3$, a very efficient particle removal process is required to achieve these limits. The size of the fine particles in the produced gas typically ranges from submicron to a few microns. Abatement of such fine particulates from the produced gas at higher temperature can be efficiently achieved by hot gas filtration. This has been well proven in many laboratory and demonstration biomass gasification plants by operating hot gas filters with sizes from 1 to 100 filter candles since the beginning of the 1990s. So far, these hot gas filters have been preferably operated at temperatures between 500°C and 600°C. Under these conditions, condensation of tar does not occur and the likelihood that particulates which typically have a high alkaline content become soft and sticky is low. Softening and sintering of particulates can occur at high temperatures. Chlorides, such as NaCl, KCl, or $CaCl_2$, decrease the softening temperature of the particulates. Additionally, in case of a eutectic mixture, the softening temperature can be decreased significantly. At high temperatures, also chemical solid phase reactions can occur which may change the properties of the dust cake. Softening and sintering of particulates results in an instable filtration, which means that the differential pressure of the filter increases continuously since the dust layer cannot be removed by applying a reverse blowback pulse. A comprehensive description of thermal property changes of dusts and the influence on the filtration at higher temperatures is given in the review on hot

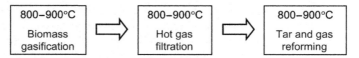

FIGURE 4.3 Concept for an energy efficient syngas cleaning process.

gas filtration by Heidenreich [27]. The interest in performing hot gas filtration of produced gas from biomass gasification at high temperatures of 800–900°C increased in recent years due to a new energy efficient gas cleaning approach, see Fig. 4.3.

The main goal of this concept is to avoid gas cool-down to perform filtration and subsequent heat-up of the gas to perform catalytic tar reforming. By performing the filtration at the gasification temperature of 800–900°C, which is also the temperature range required for effective catalytic tar reforming, the overall energy efficiency of the process is improved. Additionally, the tar reforming catalyst is efficiently protected from deactivation by particle deposition and fouling by placing the filter upstream of the catalyst unit. Even if the interest in the filtration of produced gas from biomass gasification at high temperatures of 800–900°C is high, only limited experimental test results achieved at these high temperatures are available so far. Tests performed at Delft University of Technology (the Netherlands) with a steam/oxygen blown $100\,kW_{th}$ atmospheric circulating fluidized bed gasifier and a hot gas filter downstream of the gasifier have shown first promising results that stable filtration can be achieved at 800°C [28–30]. The hot gas filter was equipped with three rigid ceramic filter candles with an outer diameter of 60 mm, an inner diameter of 40 mm, and a length of 1.5 m. Two different types of ceramic filter candles made of silicon carbide, Dia-Schumalith[1] and Dia-Schumalith[1] N filter candles of Pall Corporation, were used. Both kinds of filter candles have an asymmetric structure with a thin fine filtering membrane layer of mullite grains integrally sintered onto the outer surface of the coarse support body. The support body of both types of filter candles is made of silicon carbide grains. The silicon carbide grains of the Dia-Schumalith[1] hot gas filter element are clay bonded. The Dia-Schumalith[1] N filter elements have a different binder system, specially developed for higher temperature application, with a higher mechanical strength, a higher thermal stability, an improved corrosion resistance, and an improved creep resistance at higher temperatures compared to the clay bonded Dia-Schumalith[1] filter candles. The fine membranes achieve filtration of particles with sizes down to less than 0.3 μm. The combination of the support body and the fine filtering membrane guarantees a low differential pressure at high filtration fineness and an excellent cleaning behavior. Filter elements manufactured from silicon carbide

---

1. Dia-Schumalith is a trademark of Pall Corporation.

are well known for their high thermal shock resistance and high thermal and chemical resistance. For this reason these filter elements are well suitable for the filtration of hot gases even if fast temperature changes may occur. The filter elements exhibit very high mechanical strength. Different biomasses, A-wood, B-wood, miscanthus, and straw, were used as feedstock in the experiments and their influence on the filtration behavior was investigated. Besides the different behavior and characteristics of the ashes of the different feedstock, the tests indicated the importance of a careful selection of the regeneration parameters to achieve stable filtration. Several test series have been performed, where each test series lasted only for some 10 h. Longer tests could not be performed with the pilot unit at the university. However, the general feasibility to filter syngas from biomass gasification at 800°C has been shown. Even for biomass feedstock with high alkaline content, such as straw, stable filtration behavior could be achieved. Due to the limited test duration, additional long-term tests at high temperatures are still required to prove stable filtration as well as long-term durability of the filter material for industrial applications.

High temperatures place high demands on the mechanical, thermal, and chemical stability of the filter material that is used. Additionally, the filter media has to be stable against the chemical composition of the gas atmosphere and of the particulates. The higher the temperature of the gas is, the higher the demands on the filter material are. Flexible filter materials, applied at low temperatures, would be destroyed by the high mechanical stresses caused by the movement of the material by the reverse blowback pulses applied for the regeneration of the filter elements. At high temperatures, only rigid self-supporting ceramic or metallic filter elements are used. Metallic filter elements have some temperature limitations. Filter elements made of stainless steel can be applied for temperatures up to 420°C. For temperatures up to 650°C, filter elements made of high temperature steels can be applied. If the gas contains sulfur or chloride components, only metal filter elements made of special metal alloys can be applied. Since syngas from biomass gasification typically contains hydrogen sulfide, ceramic filter materials are the preferred option for this application. Additionally, when the gas shall be filtered at very high temperatures above 800°C, ceramic is the only option. Ceramic filter materials are available as high density and low density ceramics. High density ceramics are made of sintered grains preferably of silicon carbide, alumina, or cordierite. Low density ceramics are mostly made of aluminosilicate fibers. The porosity of high density ceramics is in the range of about 40%. Low density ceramic filter elements have a high porosity of up to about 90% and a high internal surface due to the fiber structure. They are typically vacuum-formed from fiber slurry and the fibers are bonded together at the contact points by inorganic and/or organic binders. By the vacuum-forming manufacturing process high porosity of the filter elements is achieved. Low density ceramics have high fracture toughness due to the loose structure. However, the mechanical strength is very low according to the loose structure of the fibers and the risk of candle breakages is correspondingly high.

The intensity of the back pulse for regeneration needs to be controlled so that no fibers are released from the filter structure. The differential pressure of low density ceramics is relatively low. High density ceramics have a mechanically very stable structure with a high mechanical strength. High density ceramic filter elements made of silicon carbide can have bursting pressures of higher than 5 MPa and O-ring pressure strength of more than 20 MPa. Manufacturing of high density ceramics is performed either by pressing or extrusion followed by sintering at high temperature. By selection of the right grain sizes, the pore size and the size distribution of the filter elements can be very exactly adjusted. Furthermore, the pore size and porosity can be adjusted by adding pore-forming materials which burn out during the sintering process, such as for example, polymers, sawdust, or graphite. By the amount and particle size of the pore-forming material, the pore size and the porosity can be controlled. The collection efficiency of high density ceramic filter elements is very high and reaches nearly 100% [31], and even for submicron particles this performance can be achieved [32]. Thermal stability of ceramic filter media depends on the material used. Temperatures up to 1000°C and higher are possible.

The integration of filter candles directly into the freeboard of a fluidized bed gasifier increases the requirements on the filter material further. Even if temperatures in the freeboard of the gasifier will be under normal operation conditions typically below 850°C, temperature peaks can occur during the operation of the gasifier, which requires a filter material resistant at least to 900°C, better to 950°C. Moreover, high fluid dynamic forces can act on the filter candles in the freeboard of the fluidized bed gasifier. Depending on the gas flow velocities and on the gas turbulences, attrition on the filter candle surface might occur.

Produced gas from steam biomass gasification, containing high amounts of water vapor, dust, gaseous sulfur and alkaline components, is a quite aggressive gas environment with high influence on the durability of filter elements, especially at high temperatures. Investigations on the corrosion behavior of silicon carbide filter elements under biomass gasification conditions at high temperatures of 800–950°C by Schaafhausen et al. [26] have shown that corrosion by water vapor oxidation is a special form of corrosion and a serious problem for silicon carbide materials. The water vapor oxidation caused an expansion of the silicon carbide filter material. In general, the silicon carbide material is protected by a SiO_2 layer. It is assumed that the water vapor hydrates the SiO_2 layer which accelerates the transport of the water vapor and increases the oxidation rate. Additionally, the presence of alkalines increases the oxidation rate. An alkali silicate melt can be formed on top of the SiO_2 layer which increases the water vapor diffusion through the SiO_2 layer. Thus, the oxidation of the silicon carbide material beneath the SiO_2 layer is enhanced and causes formation of gaseous CO. This leads to the expansion and dissolution of silicon carbide material. The investigations of Schaafhausen et al. showed that even if silicon carbide filter elements have been used successfully at 800°C in biomass gasification environment for short-term operations of some 10 h as shown above, silicon carbide filter elements are not the right choice for long-term operation in the freeboard of a fluidized bed biomass gasifier where temperatures significantly above 800°C can occur.

More promising ceramic materials for use as filter material in high water vapor containing biomass gasification environment are mullite and alumina. It has been shown by exposure tests of corundum- and mullite-based filter materials in a reducing, alkali-rich gas and in contact with different ashes from biomass gasification that the best solution for the application in high temperature gasification atmospheres is to use corundum-based filter materials with silica-free binder phases [25].

4.4   TAR REDUCTION

One of the main problems in improving the industrial viability of biomass gasification processes is related to the presence of tar in the product gas. Tar is a quite complex mixture of different condensable hydrocarbons including one and multiple ring aromatics as well as oxygen containing hydrocarbons. Tar is considered as all organics with a molecular weight larger than that of benzene [33]. Tar is generated during the gasification process and can cause serious problems in the process by condensation in downstream gas cleaning equipment, in pipings as well as in subsequent power generation or fuel generation processes. Tar can be eliminated from the produced gas by steam or dry reforming as well as by cracking or hydrocracking. Steam reforming is considered the most appropriate way for reduction of tar [34]. However, the reforming and cracking reactions require high temperatures, above 1200°C, to be efficient due to high activation energies, in most cases these are greater than 250–350 kJ/mol. With the use of catalysts, steam and dry reforming reactions become an effective way to remove tar components from the produced gas at lower temperatures.

Many different materials have been investigated for their catalytic tar reforming performance, among them the most important ones are nickel-based catalysts, novel metal catalysts, iron-based catalysts, dolomite, olivine, and carbon.

Generally, tar reduction processes can be applied inside the gasifier, then called primary tar reduction, or downstream outside the gasifier, then called secondary tar reduction. One of the most important advantages of a fluidized bed gasifier is that a low cost mineral bed material catalytically active for the reduction of tar can be used in presence of steam as a gasification agent [35,36]. This primary catalyst to be utilized in the gasifier as bed material should be at the same time efficient in the reforming of hydrocarbons, have high selectivity for syngas, and high resistance for attrition and coke formation. It should also satisfy the requirement of relatively low cost, because the formation of ash and char obliges to discharge used material and reintegrate the bed inventory continuously with a fresh/regenerated charge. A detailed analysis of the performance of primary tar removal catalysts applied at lab to industrial scale is beyond the scope of this book. A comprehensive review of primary tar reduction measures in biomass gasification processes (including gasification conditions and gasifier design in addition to bed material or additives) has been published by Devi and coauthors [37], and a more general and recent examination of different strategies for tar reduction in the biomass product gas via mechanical, catalytic, and thermal methods [38] also provides quantitative and comparative figures about catalytic hot gas conditioning performed within the gasifier bed itself.

Many investigations deal with dolomite, $(Ca,Mg)CO_3$, or olivine, $(Mg,Fe)_2SiO_4$. Calcined dolomite, limestone, or magnesite have been found able to increase the gas hydrogen content [39–43]. Olivine shows a slightly lower activity in biomass gasification and tar reforming, but higher attrition resistance than dolomite [44–46]. Adding calcined dolomite to the fluidized bed inventory of the gasifier allows to reduce the tar content in the dry product gas from two digit numbers down to $1–2\,g/m_N^3$, while with olivine the corresponding average value is $5–7\,g/m_N^3$. A comparison of fresh olivine and used olivine from the industrial gasifier of the Güssing plant showed higher hydrogen and carbon dioxide content and lower carbon monoxide content in the produced gas for the used olivine. Additionally, the water-gas shift (WGS) reaction was enhanced and the tar content was decreased by about 80%. The formation of a calcium-rich layer on the used olivine due to the contact and interaction with components of the biomass ash and additives has been identified as the reason for this difference [47].

Nickel-based reforming catalysts show higher activity and selectivity for tar conversion to hydrogen-rich gas, increasing noticeably the gas yield at the expense of char and tar, but when used as bed material or as an additive to the bed material suffer from (1) mechanical fragility, (2) rapid deactivation mostly due to sulfur, chlorine, alkali metals, coke, (3) metal sintering, altogether resulting in limited lifetime [20]. Additionally, Ni-based catalysts achieve very effective reduction of NH_3 in the product gas. Very positive results with respect of all points just mentioned were obtained by impregnation of olivine with nickel [48]. The mechanism of the active phase formation in Ni-olivine under biomass gasification conditions is well understood [49] and the large-scale preparation controlled. The positive features of natural olivine (mechanical resistance and activity in tar reforming) are combined with those of nickel catalysts (high activity in reforming of hydrocarbons), without the disadvantages encountered with commercial products. When the Ni-olivine catalyst was used in a pilot scale (100 kW thermal) dual fluidized bed gasifier as primary catalyst for tar destruction and methane reforming instead of pure olivine it showed very good resistance to attrition and coke formation, with an order of magnitude reduction in the tar content of the product gas [36]. The main drawback to an extensive utilization of Ni-olivine in the fluidized bed of biomass gasifiers is due to existing stringent constraints on nickel discharge, as Ni is a harmful heavy metal. Using a Ni-based bed material would contaminate the fly ash as well as the slag and would increase considerably the cost of disposal for both. Due to this fact developments were carried out to increase the catalytic activity of olivine by using nontoxic and cheaper impregnation materials than Ni. It is well known that olivine samples with similar general composition and phase structure may show different catalytic activity depending on the degree of integration of iron into their respective crystalline structure [50]. Calcination of iron-bearing olivines in air leads to the oxidation of iron and formation of iron free oxides, the presence of which affects the olivine activity. The amount of iron oxides formed is dependent on calcination time and temperature [50,51]. Recently, using an optimized impregnation method, the Fe content of olivine was enriched with an additional 10 wt% (corresponding to a total iron weight percentage of about 16 wt%), and catalytic biomass gasification experiments were performed at bench scale [52]. These bench scale tests showed an improved tar reduction by using olivine and Fe-olivine as bed material, respectively, compared to silica sand. With silica sand as bed material and a gasification temperature of 800°C, the tar content in the syngas was $16.8 \, g/m_N^3$ compared to 5.5 and $3.7 \, g/m_N^3$ with olivine and Fe-olivine, respectively. When 10 wt% Fe-olivine was tested in a pilot gasifier instead of olivine, the gas yield increased on average by 40% and the hydrogen yield by 88%. Correspondingly, the methane content in the syngas was reduced by 16% and tar production per kg of dry ash free (daf) biomass by 46% [53]. Complete characterization and microreactor reactivity data are also available for Fe-olivine materials containing up to 20 wt% of iron [54] and definitively confirm the interest in addition of iron on olivine.

When using Fe-olivine as bed material in a dual fluidized bed gasifier, Fe-olivine has a double effect on tar reduction: (1) on one hand, it acts as a catalyst for tar and hydrocarbon reforming as well as it enhances the WGS reaction; (2) on the other hand, it also acts as an oxygen carrier transporting oxygen from the combustor to the gasifier for partial oxidation of tar and hydrocarbons.

By primary tar reduction, the tar content of the produced gas can be reduced to levels of some $1-2 \, g/m_N^3$. This is a significant decrease of the tar content but far from the requirements for power generation and fuel synthesis processes. For example, tar levels of below $100 \, mg/m_N^3$ are required for gas engines [55,56] and below $50 \, mg/m_N^3$ for gas turbines [56]. In the case of an envisaged Fischer–Tropsch downstream fuel synthesis process the tar content in the syngas has to be reduced below 1 ppmv [56] to prevent deactivation of the catalyst by tar condensation. Therefore, secondary catalytic tar reduction is needed as an additional step. This secondary catalytic unit has to be protected by a hot gas filter to prevent fast catalyst deactivation by particle deposition.

An analysis of Corella et al. [57] showed that only for a tar concentration below $2 \, g/m_N^3$ can deactivation of catalysts by coke formation be avoided. For higher tar contents, the rate of coke formation on the catalysts is higher than the simultaneous gasification rate of the coke by H_2O and CO_2. Therefore, primary tar reduction in the gasifier to a level of $2 \, g/m_N^3$ would very well fit to an additional secondary tar reduction to achieve very low tar levels in the syngas.

Secondary tar reduction can be performed by using fixed bed catalysts or monolithic catalysts downstream of the gasifier. The most commonly investigated catalysts include commercial Ni catalysts applied for steam reforming, iron catalysts, char beds, and specially developed supported Ni, Rh, Ru, Pt, Pd catalysts on different support materials (eg, Al_2O_3, SiO_2, ZrO_2, MgO, MgO-CaO) and also in some cases comprising additional promotors (eg, WO_3, CeO_2). The main target of catalyst developments is to achieve high tar conversion and high resistance against coke formation and sulfur poisoning which are the main problems for catalyst deactivation related to catalytic tar reduction in biomass gasification gas. Metal catalysts have a strong interaction with their support material. For example, Ni on acidic alumina support shows high deactivation by coke formation. In contrast, Ni on a basic MgO support shows a very high resistance to coke deposition [58]. Sato and Fujimoto reported on a high resistance to coking and sulfur poisoning of a Ni catalyst promoted with WO_3 and supported on MgO-CaO [59]. Nacken et al. showed also a high tar conversion performance of Ni supported on MgO in the presence of 100 ppmv H_2S [60].

An alternative to fixed bed or monolithic catalysts downstream of the gasifier is the use of catalytic filter elements for secondary tar reduction. Catalytic filter elements are a very promising technology to combine particle and tar removal at high temperatures (see eg, Refs. [61,62]). Different designs for the integration of the catalyst into the filter elements have been developed. In Fig. 4.4A an integration of the catalyst by a catalytic coating of the inner support structure of the filter element is shown. The coating was performed by a wet impregnation process [61]. Fig. 4.4B shows an alternative approach where the catalyst is integrated into the filter element as a fixed bed accommodated in the open cylindrical space between the filter element and an additional porous tube placed coaxial in the interior of the filter element [62]. A third design is shown in Fig. 4.4C. In this approach a catalytically impregnated ceramic foam tube is added into the interior of the catalytically impregnated filter element [63].

Catalytic filter elements with a Ni catalyst based on MgO have shown in laboratory tests complete naphthalene conversion at 800°C and a filtration velocity of 90 m/h in the presence of 100 ppmv H_2S and a naphthalene concentration of 5 g/m_N^3 in the gas [60]. Long-term tests over 100 h under the same operating conditions showed no deactivation of the catalyst. Tests at pilot scale using a hot gas filter with catalytic filter elements downstream of a biomass gasifier showed a promising proof of this concept [64].

Combining particle removal and tar reforming in a catalytic filter is an interesting approach to simplify the gas cleaning process by reducing the number of required unit. In this way, the gas cleaning process becomes more energy efficient and the investment cost and space requirements are reduced.

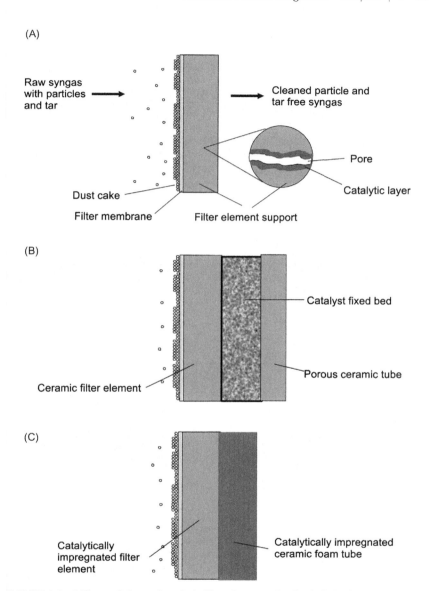

FIGURE 4.4   Different designs of catalytic filter elements. (A) Catalytically impregnated filter element. (B) Filter element with integrated catalyst fixed bed. (C) Catalytically impregnated filter element with integrated catalytically impregnated ceramic foam tube.

4.5 VALIDATION OF THE UNIQUE GASIFIER CONCEPT— RESULTS OF TAR AND DUST ABATEMENT BY CATALYTIC FILTER CANDLES IN THE FREEBOARD OF A FLUIDIZED BED GASIFIER

The innovative UNIQUE gasifier concept with the integration of catalytic filter elements into the freeboard of a fluidized bed gasifier, as describe in Section 4.1 is the next step in simplifying the biomass gasification process. The feasibility of this concept has been firstly investigated and proved by laboratory tests at real process conditions. In these bench scale tests, a catalytic filter candle with a total length of 443 mm was integrated in the freeboard of a 100 mm ID gasifier which was electrically heated. Fig. 4.5 shows a scheme of the bench scale facility used for the lab tests [65]. Fig. 4.6 shows a scheme of the gasifier with the integrated candle.

Olivine particles with a mean Sauter diameter of 0.306 mm were used as bed material. Almond shell particles with a mean Sauter diameter of 1.022 mm were fed continuously into the hot fluidized bed at a constant rate of about 4–6 g/min. Tar and particles in the produced gases are sampled according to the UNI CEN/

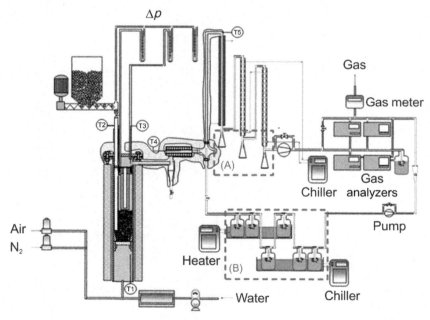

FIGURE 4.5 Scheme of the bench scale fluidized bed biomass gasification plant (cyclone and filter are bypassed when the ceramic candle is placed in the gasifier freeboard): (A) Tar in the condensate samples determined by total organic carbon (TOC) analysis; (B) Tar fraction sampled in 2-propanol, according to the Technical Specification CEN/TS 15439 and analyzed by GCMS or HPLC/UV.

TS 15439 protocol. This method is based on the discontinuous extractive sampling under isokinetic conditions of a representative part of a tar containing gas stream. The sampling train consists of a heated probe with a heated particle filter for the removal of solid matter. The volatile tars are trapped in heated or chilled impinger bottles containing a known quantity of 2-propanol. Quantitative determination was carried out by using calibration curves of pure standard tar compounds: phenol (Ph-OH), toluene (Tol), styrene (Styr) indene (Ind), naphthalene (Nap), biphenyl (Bph), diphenyl ether (DphE), fluorene (Fle), phenanthrene (Phe), anthracene (Ant), fluoranthene (Fla), and pyrene (Pyr) [66]. Online gas analyzers (IR, UV, and TCD) were used for real-time detection of H_2, CO, CO_2, CH_4, NH_3, and H_2S.

Many test runs were performed using different catalytic filter candles and with variation of the operating parameters. Fig. 4.7 shows a comparison of results obtained in some selected test runs with catalytic candles, compared to the reference test without a filter candle installed.

Al_2O_3 based catalytic filter candles were used in these selected tests. In test I, a catalytic filter candle with an MgO-Al_2O_3-based Ni catalyst coating performed by a two-step incipient wetness impregnation process was used. The catalytic

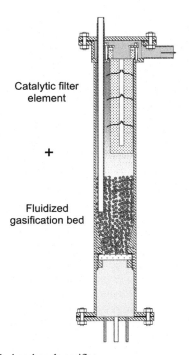

Catalytic filter
element

+

Fluidized
gasification bed

FIGURE 4.6   Scheme of the bench scale gasifier.

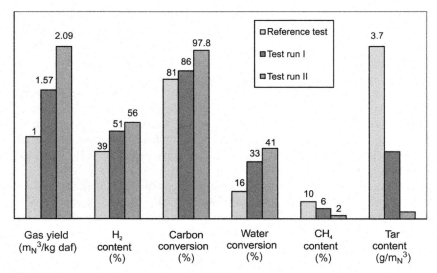

FIGURE 4.7 Results of some selected test runs with catalytic candles, compared to the reference test without a filter candle installed.

filter candle used in Test II was prepared according to the special design shown in Fig. 4.4C and contained additionally a catalytically activated hollow-cylinder ceramic foam tube which was integrated into the interior of the catalytic coated filter candle. Test I is characterized by a gasification time of 6 h, while the overall duration of Test II is 20 h (three consecutive periods called a, b, and c, with intermediate char burning steps). The catalytic filter candle used in test run II showed the highest catalytic activity. This result is as expected since this candle carries more catalyst than the catalytic candle used in test run I. The dry gas yield was improved with the use of the catalytic filter candles. With the catalytic candle used in test run II the gas yield was more than doubled compared to the reference case. In test run I, the gas yield was improved by about 60%. By using the catalytic filter candle of test run II, the carbon conversion achieved almost 98% and the water conversion obtained is quite close to the calculated equilibrium value. This is a noteworthy result, as low water conversion is often considered a drawback for steam gasification [67]. The reforming activity of the catalytic candles can be quantified by comparing the values of the measured steam conversion with the corresponding thermodynamic equilibrium values calculated at the test respective conditions for biomass steam gasification. The hydrogen content of the produced gas was improved from 39% (reference run) up to 56% in test run II. In the reference run less than 60% of the hydrogen content of the biomass was converted to H_2 in the product gas, while in each test with the catalytic candle this value greatly exceeded 100% as a result of enhanced hydrocarbon reforming reactions with steam. The methane and tar content in the

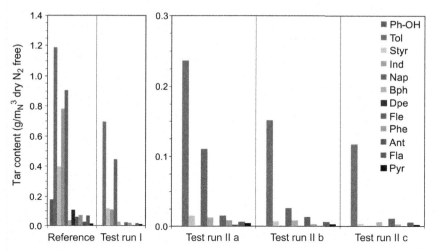

FIGURE 4.8 Characterization of tar samples of the reference test as well as of test run I and II. *Results reproduced from Rapagnà S, Gallucci K, Di Marcello M, Foscolo PU, Nacken M, Heidenreich S, et al. First Al₂O₃ based catalytic filter candles operating in the fluidized bed gasifier freeboard. Fuel 2012;97:718–24.*

produced gas was decreased by using the catalytic filter candles. Methane was reduced from a content of 10% in the reference run to 6% in test run I and to 2% in test run II. Tar was reduced from 3.7 to 1.47 g/m$_N^3$ in test run I and to 0.15 g/m$_N^3$ in test run II which corresponds to tar conversions of 60% in test run I and 96% in test run II, respectively. Methane and tar reduction indicates that the integrated higher amount of basic oxide supported Ni catalyst in the catalytic candle used in test run II results in a very high reforming activity. Fig. 4.8 shows the characterization of the tar samples. Among different tar species, toluene appears as the prevailing compound after the catalytic reforming. In the catalytic candle, methane and tar steam reforming takes place together with WGS. From the experimental results and thermodynamic calculations it is estimated that there was a corresponding thermal energy demand of the order 0.5 MJ/kg of biomass. Computational Fluid-Dynamics (CFD) simulations to characterize the behavior of a catalytic filter candle in the freeboard of a fluidized bed gasifier confirm that the temperature drop expected through the candle is limited [68]. The enhancement of the reforming reactions inside the gasifier vessel allows to optimize the supply of this thermal load, provided by gasification itself, with respect to alternative process layouts with downstream equipment for gas conditioning treatments, where energy from additional sources is often required [69].

Without taking into account the low molecular weight hydrocarbon toluene, very high tar conversion was achieved as shown in Fig. 4.9.

In test run I, the tar concentration in the cleaned syngas was 0.77 g/m$_N^3$ without toluene and the corresponding tar conversion without toluene was

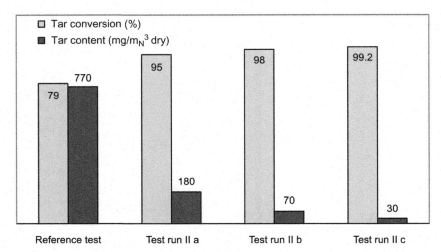

FIGURE 4.9 Tar conversion and tar content calculated without taking into account the low molecular weight hydrocarbon toluene.

79% compared to 60% with toluene. In test run II, the lowest tar concentration achieved was $30 \, \text{mg/m}_N^3$ without toluene and the corresponding tar conversion without toluene was 99.2% compared to 96% with toluene. The results indicate that the catalyst in test run II is very active even after 20 h of operation without any deactivation.

Total carbon in the permanent gaseous phase per kg of biomass was increased on average by 30% as a result of tar conversion in smaller molecules. This substantial increase in the gas yield is partially counterbalanced by a reduction in the lower heating value of the fuel gas by 13–16%.

The initial differential pressure of the catalytic filter candle in test run I was about 1.7 kPa and increased constantly to a final value of 2.85 kPa after 3 h of gasification. In test run II, the differential pressure of the catalytic candle increased from 2.25 to 3.0 kPa, remaining constant at this level until the end of the gasification test suggesting that the thickness of the powder cake tends to a stable value. The differential pressure values are in an acceptable range for this kind of application. In the test runs, no periodic blow back for regeneration of the filter candles has been applied like it is in the case of industrial applications.

An industrial-scale benchmark of the effectiveness of the UNIQUE gasifier concept was obtained by testing a catalytic filter candle inside a commercial gasifier [70]. The testing was performed at the Güssing plant (8 MW$_{th}$). The filter candle was installed in the freeboard of the gasifier as shown in the sketch of Fig. 4.10.

A slip stream was taken and the filter candle was periodically cleaned from the dust cake by means of a back pulsing system for rigid hot gas filter elements (see Fig. 4.11).

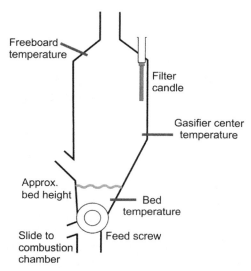

FIGURE 4.10   Scheme of filter test module installed in the freeboard of the Güssing gasifier.

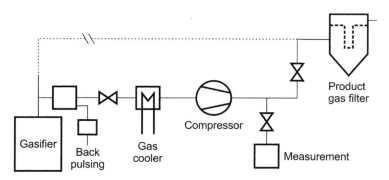

FIGURE 4.11   Slip stream set-up.

Nitrogen is used as the blowback gas for regeneration, properly preheated to avoid possible condensation problems. Due to high operating temperatures of ≥850°C in the freeboard of the Güssing gasifier, a newly developed corundum-based catalytic filter candle of commercial length (1.5 m) installed into the freeboard of the gasifier showed promising results. The test showed that the particulate content of the cleaned gas is practically brought to zero by the filter candle, making it compatible with the most sophisticated applications, for example, fuel cells. Stable filtration of raw product gas was accomplished. The particulate concentration of the raw gas in the Güssing gasifier was 56.3 g/m$_N^3$ during the test. Tar reduction was as high as 95%. The nickel-based catalytic filter candle of full commercial length could be used successfully in integrated high temperature reforming of tar and removal of particles from biomass gasification product.

A further proof of concept has been realized at the Trisaia research centre of ENEA (Italian National Agency for Renewable Energy and the Environment) in the framework of the European R&D Contract UNIfHY [5]. An innovative atmospheric steam/oxygen gasifier of 1 MW thermal load (the principle design of this gasifier is described in Foscolo et al. [71]) was modified according to the UNIQUE gasifier concept for the installation of 60 filter candles in the freeboard of the fluidized biomass gasifier. For continuous operation, a special blowback system was also integrated. Corundum-based noncatalytic filter candles were installed very recently to prove the concept on full-scale size for the first time. First, so far unpublished, results demonstrate a very successful continuous and stable filtration and operation of the gasifier. Results will be published in detail in the near future [71a]. A study on high temperature (800°C) filtration coupled to a lab scale gasifier showed the additional benefit of a prereforming effect: the total amount of tars in the gas was reduced on the noncatalytic filter in all tests regardless of the used feedstock, bed material, or gasifying agent, in the order of up to 50% [72].

As a whole, feasibility of the concept is proven and integration and operation of filter candles in the gasifier freeboard appears to be a very promising option for hot gas cleaning toward tar and particulate.

4.6 ABATEMENT OF DETRIMENTAL INORGANIC TRACE ELEMENTS

Studies were performed on high temperature removal of the most relevant trace elements, namely alkalis and sour gases like H_2S and HCl, by means of appropriate sorbents to be added to the fluidized bed inventory of the UNIQUE gasifier [11,73,74]. A thermodynamic process model assuming Gibbs free energy minimization was used to investigate the fate and removal of sour gases and alkali species in the biomass gasification process, as a function of different composition of inlet streams and gasifier temperature. To confirm theoretical findings, experimental tests were performed on conventional and new synthetic solid sorbents suitable to be utilized at the operating conditions of the gasifier and able to assure very low concentration levels of these contaminants in the gas phase, comparable with threshold values recommended for SOFCs and production of biofuels by chemical syntheses.

Contaminants are typically removed in downstream equipment. Adding suitable sorbents to the fluidized bed inventory of the gasifier (see Fig. 4.1) allows for additional beneficial effects: the prevention of bed agglomeration caused by alkali compounds, reduction of corrosion of the gasifier walls by sour gases, and improvement of catalyst activity for tar reforming in the filtering elements placed in the freeboard of the gasification reactor by minimization of poisoning by trace elements.

An issue is that of separating the sorbents from the remaining constituents of the gasifier bed inventory, with the aim to perform regeneration and cyclic utilization of them: classification based on particle size and density appears as an appropriate means. However, this has not been investigated so far.

4.6.1   Alkali Species

Alkali species cause bed agglomeration, fouling, and corrosion. As a result of the relatively high potassium and chlorine content of biomass, KCl is the most important compound released during biomass gasification. Several investigations show that aluminosilicates are highly suitable for alkali removal [75–77]. Investigations on the sorption mechanism show the necessity of water [78–81], as can be seen in reaction (4.1). Attention is often focused on the analysis of the sorbent capacity more than on the gas purity downstream of the sorbent, which is the most important figure, the latter being rarely determined in a direct way. It was determined either indirectly by analyzing the sorbent capacity and back calculation of the gas purity or by condensing the gas stream [82–84]. Under gasification conditions aluminosilicates reduced the alkali concentration to the ppb-level [85]. In addition, aluminosilicates rich in alumina remove other trace metals like zinc [86] or arsenic [87].

$$Al_2O_3 \cdot xSiO_{2,(s)} + 2AlkCl_{(g)} + H_2O \leftrightarrow 2AlkAlO_2 \cdot xSiO_{2,(s)} + 2HCl_{(g)} \tag{4.1}$$

The results of the thermodynamic calculations of alkali sorption on aluminosilicates (Fig. 4.12) show that the alkali chloride concentration in the producer gases can be limited to values of several ppbv to 4 ppmv depending on the biomass type [11]. The concentration of alkali hydroxides is about two orders of magnitude lower. The observed biomass influence is caused by the different HCl concentrations in the producer gas, which result either from the direct HCl release during gasification or from indirect HCl release during alkali chloride sorption according to reaction (4.1). Anyway, the biomasses which contain the highest chlorine amounts lead to the lowest alkali chloride sorption performance. Although KCl is the most abundant of all alkali species in the raw producer gas, NaCl is the alkali species with the highest concentration in the cleaned gas. Thus, the condensation temperature decrease depends on the residual NaCl concentration in the cleaned gas. This ranges from 530°C to 630°C, which is far below its melting point resulting in a drastically reduced risk of fouling and corrosion downstream the gasifier, specifically when a SOFC is operated (at temperatures of about 800°C). The predicted potassium-containing aluminosilicate phase is leucite ($KAlSi_2O_6$).

The sorption experiments reported by Stemmler et al. [74] were performed as packed bed experiments at atmospheric pressure. The inlet gas stream was laden with KCl by overflowing a KCl source resulting in a KCl gas concentration of about 20 ppmv. At a space velocity of 9800 h^{-1} no evidence of kinetic limitation was observed. The results of the KCl sorption tests confirm that

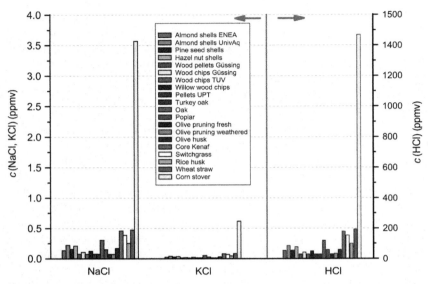

FIGURE 4.12 Calculated remaining NaCl and KCl concentrations in the cleaned syngas and parallel occurring HCl concentrations [3].

aluminosilicates like bauxite, kaolinite, and naturally occurring zeolite are suitable sorbents for KCl removal below 100 ppbv at 800–900°C (Fig. 4.13). A reason for the unsuitability of bentonite might be the layer structure of montmorillonite, the main constituent of bentonite [88]. In this structure, the Al_2O_3 is covered by SiO_2 and, thus, cannot easily participate in the sorption reaction. However, both oxides, SiO_2 and Al_2O_3, need to take part in the sorption reaction to achieve low alkali concentrations. Since bauxite performs nearly as well as kaolinite and the naturally occurring zeolite, the higher Al_2O_3 content seems to have a minor influence. However, kaolinite is the only sorbent that keeps the 100 ppbv limit longer than 50 h. X-ray diffraction (XRD) analysis confirms that microcline ($KAlSi_3O_8$), which was already predicted by the thermodynamic calculations, is formed as potassium-containing aluminosilicate phase.

Fluidized bed gasification tests at atmospheric pressure in presence of bauxite for the removal of alkali chlorides, that is, KCl (and NaCl), confirmed the suitability of aluminosilicate sorbents for the removal of alkalis [89]. The used aluminosilicates are easily available, cheap, and without environmental implications for their disposal after utilization.

4.6.2 H_2S

H_2S is the most released sulfur species [90]. The H_2S sorption mechanism using metal oxides as shown in reaction (4.2) or metals as shown in reaction (4.3) and the working temperature of different sorbents have been well known for a long time [91,92].

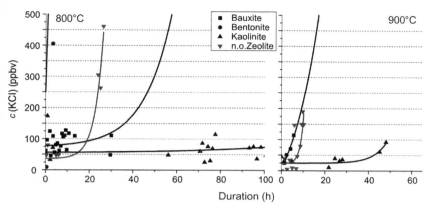

FIGURE 4.13   KCl concentrations in the gas leaving the sorbent bed [2].

$$MeO_{(s)} + H_2S_{(g)} \leftrightarrow MeS_{(s)} + H_2O_{(g)} \qquad (4.2)$$

$$Me_{(s)} + H_2S_{(g)} \leftrightarrow MeS_{(s)} + H_{2,(g)} \qquad (4.3)$$

Calcium based sorbents have been recognized for a long time as effective material for H_2S removal at high temperature. The utilization of calcined and uncalcined limestone has been studied [93,94]. Even though the sorption reaction already shows its sensitivity to water, as given in reaction (4.2), or hydrogen, as given in reaction (4.3), many gases used for investigations were balanced with nitrogen or helium [95–98]. Investigations conducted in nonbalanced producer gases with steam contents of 5 vol. % are subjected to significantly different conditions [99,100]. Experimental work done at real conditions or with simulated coal gas containing high concentration of H_2S has led to the conclusion that fuel gas composition does not have a large influence on the desulfurization capacity. However, thermodynamic limitations, especially in presence of a considerable content of steam in the gas phase (see reaction 4.2), can hardly allow reaching H_2S concentrations as low as those required by SOFC.

To show the magnitude of the influence of the main producer gas constituents and temperature on the achievable H_2S concentration, unique parameters such as H_2/C ratio and ROC are used in Fig. 4.14 [11]. The lowest achievable H_2S concentrations using CaO are slightly below 100 ppmv. However, this is only a theoretical value and not of practical use, since almost no gasification agent has to be used. Therefore, the target value of 1 ppmv H_2S is not at all achievable using CaO at 850°C which has also been confirmed by experimental investigations, as shown in Fig. 4.16 [74].

Alternative systems known from literature are all characterized by drawbacks of different nature, spanning from reduction of sorption capacity with

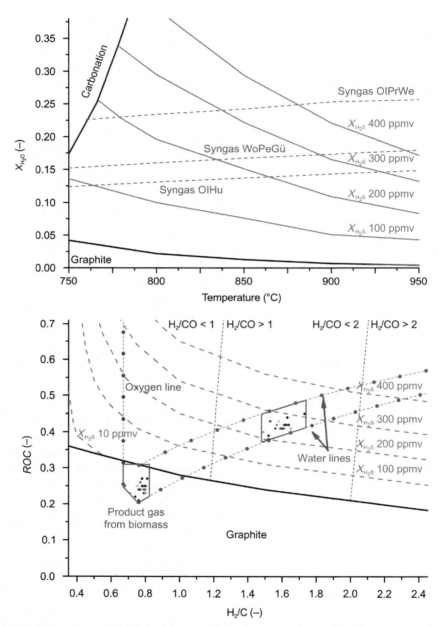

FIGURE 4.14 Achievable H_2S concentrations after sulfur absorption by CaO depending on syngas composition at 850°C (top) and temperature (bottom).

temperature (iron oxide), metal vaporization (zinc oxide), oxide reduction by H_2 and CO (copper oxide) [101]. Furthermore, they are not capable of meeting the target required by SOFC [11,74].

To meet the requirement of 1 ppmv H_2S, a new sorbent was developed according to thermodynamic predictions indicating that stabilized, Ba-based sorbents should be effective [11]. Since BaO and SrO form carbonates under "UNIQUE gasifier" conditions, the pure oxides are not suitable for H_2S reduction. However, stabilized materials that do not form carbonates should perform better. Therefore, calculations with SrO and BaO were performed excluding carbonate phases from the database. According to these calculations, the target value of 1 ppmv H_2S can be achieved with stabilized Ba-based sorbents (Fig. 4.15); Sr-based sorbents achieve only values below 10 ppmv [11].

The "CaBa" sorbent was prepared from a mixture of 10 mol. % $BaCO_3$ and 90 mol. % $CaCO_3$. In packed bed experiments, the sorbent allowed achieving H_2S concentrations lower than 0.5 ppmv (the detection limit in these experiments) in the temperature range of 800–900°C (Fig. 4.16 [74]). The stabilization effect was confirmed by XRD analysis showing the occurrence of a BaS phase. At temperatures below 760°C the H_2S concentration in the effluent stream rises up to 175 ppmv due to carbonation of the sorbent. Thus, the CaBa sorbent can be regenerated by cooling the saturated sorbent under nonoxidizing atmosphere.

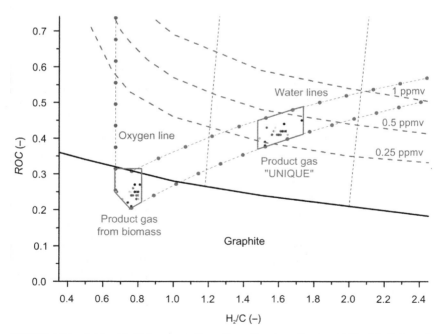

FIGURE 4.15 Achievable H_2S concentrations after sulfur absorption by stabilized BaO depending on syngas composition at 850°C.

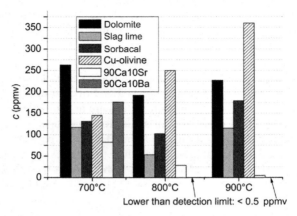

FIGURE 4.16 H_2S concentration in the gaseous stream leaving the sorbent bed.

Fluidized bed gasification tests at pilot scale ($100\,kW_{th}$) carried out at Vienna University of Technology with the CaBa sorbent confirmed the tendency toward lowered H_2S content in the product gas. However, such tests show only qualitative trends; the operating conditions, like sorbent particle size and amount charged into the gasifier, still need to be optimized.

4.6.3 HCl

Several investigations in the temperature range of 400–650°C indicate a high conversion rate for HCl with Ca-based, Na-based, and mixed Ca–Na sorbents [102–104]. The HCl sorption mechanisms are shown in reactions (4.4) and (4.5). However, these investigations do not include the achievable HCl reduction in the syngas downstream from the sorbent. Stemmler et al. [105] showed that Na_2CO_3, K_2CO_3, and an alkali rich biomass ash are suitable to reduce HCl below 1 ppmv at temperatures below 550°C in case of sufficient space velocities. Duo et al. [106] showed that the CO_2 concentration in syngas affects the behavior of HCl sorption on Ca-based sorbents but does not influence the performance of Na-based sorbents. The achievable HCl concentration is below 1 ppmv for both sorbents. In addition, this and further investigations show that a steam concentration of up to 28 vol. % does not affect the achievable HCl reduction [107]. However, all investigations were performed at temperatures much lower than the temperature of the hot gas cleaning in the "UNIQUE" process.

$$CaO_{(s)} + 2HCl_{(g)} \leftrightarrow CaCl_{2,(s)} + H_2O_{(g)} \tag{4.4}$$

$$Na_2CO_{3,(s)} + 2HCl_{(g)} \leftrightarrow 2NaCl_{(s)} + H_2O_{(g)} + CO_{2,(g)} \tag{4.5}$$

The sorption experiments at Jülich showed that the CaBa sorbent is able to reduce the HCl gas content below 0.5 ppmv over the temperature range 800–900°C [74]. This value is even lower than the value predicted by thermodynamic calculations. The formation of $BaCl_2$ phases is confirmed by XRD.

4.7 CONCLUSIONS AND OUTLOOK ON FURTHER INTEGRATION OPTIONS

Besides the deep involvement of the authors of this book in the invention and development of the UNIQUE gasifier concept, which is one reason that this concept has been selected to be described in some detail within the book, it has also been selected since it is a nice example of the benefit and challenges of process integration. By integration of process steps, the chance for a simple optimization of the single steps is replaced by a more complex adjustment of the different steps with the need to take into account the influence and dependence of the different steps on each other, respectively. The total process has to be considered and understood. Improvement of gas yield and gas composition, removal of tar, particulates, and gaseous trace elements in the UNIQUE concept is achieved by an adapted composition of the bed material, by high temperature filtration and catalytic tar and gas reforming, and by a proper selection of the operating conditions.

Further integration options in the UNIQUE concept are seen in the additional integration of membranes in the gasifier system:

1. Ionic transport membranes for selective transport of oxygen as gasifying agent into the gasifier.
2. Hydrogen separation membranes for a selective separation of a hydrogen stream from the produced gas.

The integration of ionic oxygen transport membranes into the fluidized bed gasifier could be a way for an efficient and economic operation of oxygen or oxygen-steam based gasification for small to medium scale units. Ion transport membranes are typically made of inorganic crystalline metal oxides permeable to oxygen ions at high temperatures. Typically, a temperature above 700°C is required for the transport of the oxygen ions through the dense ceramic membranes. The most common membrane materials are perovskite, fluorite, and mixtures of two materials taking advantage of the best of each material. The membranes are very thin layers supported on a porous ceramic carrier. By integrating oxygen transport membranes into the fluidized bed gasifier where the temperature is at 800–900°C, the membrane is at the right temperature to achieve a good oxygen transport. Ionic oxygen transport membranes offer an oxygen supply which requires significantly less energy than any other currently available technique for oxygen production. The general feasibility to integrate ionic oxygen transport membranes into a fluidized bed gasifier has already been investigated by some modeling work on a laboratory fluidized bed gasifier with 10 cm inner diameter [108,109]. Results showed that the membrane surface which is required to provide the necessary oxygen flow for the gasification reaction is small enough to be installed in the available space inside the gasifier bed. Positive effects which enhance the oxygen flux through the membrane are: (1) Oxygen consumption by the gasification reactions inside the gasifier increases the driving oxygen partial pressure across the membrane and thus the flux. (2) An increase in temperature by the high rate of combustion close to the surface of the membrane where the oxygen enters into the gasifier results in an increased flux since the flux depends on the absolute temperature by means of an Arrhenius-type exponential law. However, there is also the risk of local hot spots due to this enhancement.

The integration of ionic oxygen transport membranes into the fluidized gasifier bed requires very robust and stable membranes. The membrane as well as the support has to be stable in the gasification atmosphere containing high amounts of hydrogen, carbon oxides, and steam at high temperatures and they have to be resistant to the fluid dynamic forces and stresses in the fluidized bed. Especially, thermal stresses by temperature gradients are of great concern. Moreover, abrasion of the membrane layer in the fluidized bed is of concern. The investigation of the resistance and long-term durability of oxygen transport membranes in the reactive environment of the fluidized bed gasifier has to be the main focus for future experimental investigations of this concept. Additionally, sealing and fixation of the oxygen transport membrane inside the gasifier are also engineering issues, especially for larger gasifiers, which have to be considered in future investigations and developments.

A further option which is just a rough idea at the current state might be the additional integration of hydrogen separation membranes to separate a pure hydrogen stream from the produced gas. Moreover, by integration of hydrogen separation membranes to separate a hydrogen stream from the produced gas, downstream WGS reaction could be enhanced to increase the overall hydrogen yield (see Section 5.3). Hydrogen separation membranes could be integrated into the freeboard of the gasifier downstream of the filter elements or even with a special design they could be integrated directly into the interior of the filter elements. The high temperature at these positions would require the use of high temperature stable membranes, such as dense metallic or ceramic membranes.

REFERENCES

[1] Pröll T, Rauch R, Aichernig C, Hofbauer H. Performance characteristics of an 8 MW_{th} combined heat and power plant based on dual fluidized bed steam gasification of solid biomass. In: Berruti, F, (Tony) Bi X, Pugsley T, editors. Proc. Int. Conf. Fluidization XII, May 13–17, Vancouver (Canada); 2007. p. 937–44.

[2] Heidenreich S, Foscolo PU, Nacken M, Rapagnà S. Gasification apparatus and method for generating syngas from gasifiable feedstock material. PCT Patent Application PCT/EP2008/003523; 2008.

[3] Heidenreich S, Foscolo PU, Nacken M, Rapagnà S. Gasification apparatus and method for generating syngas from gasifiable feedstock material. US patent 8,562,701 B2; 2013.

[4] Foscolo PU, Gallucci K. Integration of particulate abatement, removal of trace elements and tar reforming in one biomass steam gasification reactor yielding high purity syngas for efficient CHP and power plants. In: 16th European biomass conference and exhibition 2008; Valencia, Spain, paper OA7.1.

[5] <http://www.unifhy.eu> [accessed 04.02.16].

[6] Björkman E, Strömberg B. Release of chlorine from biomass at pyrolysis and gasification conditions. Energy Fuels 1997;11:1026–32.

[7] Olsson JG, Pettersson BC. Alkali metal emission from filter ash and fluidized bed material from pfb gasification of biomass. Energy Fuels 1998;12:626–30.

[8] Zevenhoven R, Kilpinen P. Control of pollutants in flue gases and fuel gases, 2nd ed. Espoo (Finland): Helsinki University of Technology; 2002.

[9] Porbatzki D, Stemmler M, Müller M. Release of inorganic trace elements during gasification of wood, straw, and miscanthus. Biomass Bioenergy 2011;35:S79–86.

[10] Marschner H. Mineral nutrition of higher plants, 2nd ed. Amsterdam, Boston, Heidelberg, London, New York, Oxford, Paris, San Diego, San Francisco, Singapore, Sidney, Tokyo: Academic Press; 1995.

[11] Stemmler M, Tamburro A, Müller M. Thermodynamic modelling of fate and removal of alkali species and sour gases from biomass gasification for production of biofuels. Biomass Conf Bioref 2013;3:187–98.

[12] Kuramochi H, Wu W, Kawamoto K. Prediction of the behaviors of H_2S and HCl during gasification of selected residual biomass fuels by equilibrium calculation. Fuel 2005;84:377–87.

[13] Turn SQ. Chemical equilibrium prediction of potassium, sodium, and chlorine concentrations in the product gas from biomass gasification. Ind Eng Chem Res 2007;46:8928–37.

[14] Valerio V, Villone A, Nanna F, Barisano D. Biomass selection for the test campaign at the UNIQUE pilot plant. In: Proceedings of the ICPS 09, international conference on polygeneration strategies, September 1–4. Wien, Austria; 2009.

[15] Kofstad P. High temperature corrosion. London, New York: Elsevier Applied Science; 1988.

[16] Lai GY. High temperature corrosion and materials applications. Ohio: ASM International, Materials Park; 2007.

[17] Young DJ. High temperature oxidation and corrosion of metals. Amsterdam (The Netherlands): Elsevier Science; 2008.

[18] Wegge S, Grabke HJ. Einflüsse von Silicium und Kohlenstoff auf die Sulfidierung von Eisen. Mater Corros/Werkstoffe Korrosion 1992;43:437–46.

[19] Bartholomew CH. Mechanism of catalyst deactivation. Appl Catal A 2001;212:17–60.

[20] Bain RL, Dayton DC, Carpenter DL, Czernik SR, Feik CJ, French RJ, et al. Evaluation of catalyst deactivation during catalytic steam reforming of biomass-derived syngas. Ind Eng Chem Res 2005;44:7945–56.

[21] Ashrafi M, Pfeifer C, Pröll T, Hofbauer H. Experimental study of model biogas catalytic steam reforming: 2. Impact of sulfur on the deactivation and regeneration of Ni-based catalysts. Energy Fuels 2008;22:4190–5.

[22] Lisi L, Lasorella G, Malloggi S, Russo G. Single and combined deactivating effect of alkali metals and HCl on commercial SCR catalysts. Appl Catal B 2004;50:251–8.

[23] Leibold H, Hornung A, Seifert H. HTHP syngas cleaning concept of two stage biomass gasification for FT synthesis. Powder Technol 2008;180:265–70.

[24] Nielsen HP, Baxter LL, Sclippab G, Morey C, Frandsen FJ, Dam-Johansen K. Deposition of Potassium salts on heat transfer surfaces in straw-fired boilers: a pilot-scale study. Fuel 2000;79:131–9.

[25] Schaafhausen S, Yazhenskikh E, Walch A, Heidenreich S, Müller M. Corrosion of alumina and mullite hot gas filter candles in gasification environment. J Eur Ceram Soc 2013;33:3301–12.

[26] Schaafhausen S, Yazhenskikh E, Heidenreich S, Müller M. Corrosion of silicon carbide hot gas filter candles in gasification environment. J Eur Ceram Soc 2014;34:575–88.

[27] Heidenreich S. Hot gas filtration: a review. Fuel 2013;104:83–94.

[28] Simeone E, Siedlecki M, Nacken M, Heidenreich S, de Jong W. Fuel 2013;108:99–111.

[29] Simeone E, Siedlecki M, Nacken M, Heidenreich S, de Jong W. Biomass Bioenergy 2011;35:87–104.

[30] Heidenreich S, Simeone E, Haag W, Nacken M, de Jong W, Salinger M. Gefahrstoffe— Reinhaltung der Luft 2011;71:281–5.

[31] Cummer KR, Brown RC. Biomass Bioenergy 2002;23:113–28.

[32] Heidenreich S, Scheibner B. Filtration+Separation 2002;May:22–5.

[33] Maniatis K, Beenackers AACM. Introduction: tar protocols, IEA gasification tasks. Biomass Bioenergy 2000;18:1–4.

[34] Milne TA, Abatzoglou N, Evans RJ. Biomass gasification "tars": their nature, formation and conversion. National Renewable Energy Laboratory (NREL) Technical Report 1998; Golden, CO, Report NREL/ TP 570-25357.

[35] Magrini-Bair KA, Czernik S, French R, Parent YO, Chornet E, Dayton DC, et al. Fluidizable reforming catalyst development for conditioning biomass-derived syngas. Appl Catal A 2007;318:199–206.

[36] Pfeifer C, Hofbauer H, Rauch R. In-bed catalytic tar reduction in a dual fluidized bed biomass steam gasifier. Ind Eng Chem Res 2004;43:1634–40.

[37] Devi L, Ptasinski KJ, Janssen FJJG. A review of the primary measures for tar elimination in biomass gasification processes. Biomass Bioenergy 2003;24(2):125–40.

[38] Anis S, Zainal ZA. Tar reduction in biomass producer gas via mechanical, catalytic and thermal methods: a review. Renew Sustain Energy Rev 2011;15:2355–77.

[39] Delgado J, Aznar MP, Corella J. Biomass gasification with steam in a fluidized bed: effectiveness of CaO, MgO and CaO-MgO for hot raw gas cleaning. Ind Eng Chem Res 1997;36:1535–43.

[40] Olivares A, Aznar MP, Cabballero MA, Gil J, Francés E, Corella J. Biomass gasification: produced gas upgrading by in-bed use of dolomite. Ind Eng Chem Res 1997;36:5220–6.

[41] Rapagná S, Jand N, Foscolo PU. Catalytic gasification of biomass to produce hydrogen rich gas. Int J Hydrogen Energy 1998;23:551–7.

[42] Orio A, Corella J, Narvaez I. Performance of different dolomites on hot raw gas cleaning from biomass gasification with air. Ind Eng Chem Res 1997;36:3800–8.

[43] Simell PA, Hirvensalo EK, Smolander VT, Krause AOI. Steam reforming of gasification gas tar over dolomite with benzene as a model compound. Ind Eng Chem Res 1999;38:1250–7.

[44] Rapagnà S, Jand N, Kiennemann A, Foscolo PU. Steam-gasification of biomass in a fluidised-bed of olivine particles. Biomass Bioenergy 2000;19:187–97.

[45] Corella J, Toledo JM, Padilla R. Olivine or dolomite as in-bed additive in biomass gasification with air in a fluidized bed: which is better? Energy Fuels 2004;18:713–20.

[46] Devi L, Ptasinski KJ, Janssen FJJG, van Paasen SVB, Bergman PCA, Kiel JHA. Catalytic decomposition of biomass tars: use of dolomite and untreated olivine. Renew Energy 2005;30:565–87.

[47] Kirnbauer F, Wilk V, Kitzler H, Kern S, Hofbauer H. The positive effects of bed material coating on tar reduction in a dual fluidized bed gasifier. Fuel 2012;95:553–62.

[48] Świerczyński D, Courson C, Kiennemann A. Study of steam reforming of toluene used as model compound of tar produced by biomass gasification. Chem Eng Process 2008;47:508–13.

[49] Świerczyński D, Courson C, Bedel L, Kiennemann A, Guille J. Characterisation of Ni-Fe/MgO/Olivine catalyst for fluidised bed steam gasification of biomass. Chem Mater 2006;18/17:4025–32.

[50] Rauch R, Pfeifer C, Bosch K, Hofbauer H, Świerczyński D, Courson C, et al. Comparison of different olivine's for biomass steam gasificationBridgwater A.V.Boocock DGB, editors. Science in thermal and chemical biomass conversion, vol. 1. United Kingdom: CPL Press; 2006. p. 799–809.

[51] Fredriksson HOA, Lancee RJ, Thüne PC, Veringa HJ, Niemantsverdriet JW. Olivine as tar removal catalyst in biomass gasification: catalyst dynamics under model conditions. Appl Catal B 2013;130-131:168–77.

[52] Virginie M, Adanez J, Courson C, de Diego LF, Garcia-Labiano F, Niznansky D, et al. Effect of Fe–olivine on the tar content during biomass gasification in a dual fluidized bed. Appl Catal B 2012;121-122:214–22.

[53] Rapagnà S, Virginie M, Gallucci K, Courson C, Di Marcello M, Kiennemann A, et al. Fe/olivine catalyst for biomass steam gasification: preparation, characterization and testing at real process conditions. Catal Today 2011;176:163–8.

[54] Virginie M, Courson C, Niznansky D, Chaoui N, Kiennemann A. Characterization and reactivity in toluene reforming of a Fe/olivine catalyst designed for gas cleanup in biomass gasification. Appl Catal B 2010;101:90–100.

[55] Abu El-Rub Z, Bremer EA, Brem G. Review of catalysts for tar elimination in biomass gasification processes. Ind Eng Chem Res 2004;43:6911–9.

[56] Richardson Y, Blin J, Julbe A. A short overview on purification and conditioning of syngas produced by biomass gasification: catalytic strategies, process intensification and new concepts. Prog Energy Combust Sci 2012;38:765–81.

[57] Corella J, Toledo JM, Molina G. Calculation of the conditions to get less than $2\,g\ tar/m_N^3$ in a fluidized bed biomass gasifier. Fuel Process Technol 2006;87:841–6.

[58] Kong M, Fei J, Wang S, Lu W, Zheng X. Influence of supports on catalytic behavior of nickel catalysts in carbon dioxide reforming of toluene as a model compound of tar from biomass gasification. Bioresour Technol 2011;102:2004–8.

[59] Sato K. Fujimoto. Development of new nickel based catalyst for tar reforming with superior resistance to sulfur poisoning and coking in biomass gasification. Catal Commun 2007;8:1697–701.

[60] Nacken M, Ma L, Heidenreich S, Verpoort F, Baron GV. Development of a catalytic ceramic foam for efficient tar reforming of a catalytic filter for hot gas cleaning of biomass-derived syngas. Appl Catal B 2012;125:111–9.

[61] Nacken M, Ma L, Heidenreich S, Baron GV. Performance of a catalytically activated ceramic hot gas filter for catalytic tar removal from biomass gasification gas. Appl Catal B 2009;88:292–8.

[62] Nacken M, Ma L, Heidenreich S, Baron GV. Catalytic activity in naphthalene reforming of two types of catalytic filters for hot gas cleaning of biomass-derived syngas. Ind Eng Chem Res 2010;49:5536–42.

[63] Nacken M, Baron GV, Heidenreich S, Rapagnà S, D'Orazio A, Gallucci K, et al. New DeTar catalytic filter with integrated catalytic ceramic foam: catalytic activity under model and real bio syngas conditions. Fuel Process Technol 2015;134:98–106.

[64] Fantini M, Nacken M, Heidenreich S, Siedlecki M, Fornasari G, Benito P, et al. Bagasse gasification in a 100 kWth steam-oxygen blown circulating fluidized bed gasifier with catalytic and non-catalytic upgrading of the syngas using ceramic filters. WIT Trans Ecol Environ 2014;190:1079–90.

[65] Rapagnà S, Gallucci K, Di Marcello M, Foscolo PU, Nacken M, Heidenreich S. In situ catalytic ceramic candle filtration for tar reforming and particulate abatement in a fluidized-bed biomass gasifier. Energy Fuels 2009;23:3804–9.

[66] Rapagnà S, Gallucci K, Di Marcello M, Foscolo PU, Nacken M, Heidenreich S, et al. First Al_2O_3 based catalytic filter candles operating in the fluidized bed gasifier freeboard. Fuel 2012;97:718–24.

[67] Rapagnà S, Gallucci K, Di Marcello M, Matt M, Nacken M, Heidenreich S, et al. Gas cleaning, gas conditioning and tar abatement by means of a catalytic filter candle in a biomass fluidized-bed gasifier. Bioresour Technol 2010;101:7134–41.

[68] Savuto E, Di Carlo A, Bocci E, D'Orazio A, Villarini M, Carlini M, et al. Development of a CFD model for the simulation of tar and methane steam reforming through a ceramic catalytic filter. Int J Hydrogen Energy 2015;40:7991–8004.

[69] Vivanpatarakij S, Assabumrungrat S. Thermodynamic analysis of combined unit of biomass gasifier and tar steam reformer for hydrogen production and tar removal. Int J Hydrogen Energy 2013;38:3930–6.

[70] Foscolo PU. The Unique project—integration of gasifier with gas cleaning and conditioning system. In: Int. seminar on gasification 2012, Stockholm, Sweden. <http://www.sgc.se/gasification2012/> [accessed 01.02.16].

[71] Foscolo PU, Germanà A, Jand N, Rapagnà S. Design and cold model testing of a biomass gasifier consisting of two interconnected fluidized beds. Powder Technol 2007;173:179–88.

[71a] D. Barisano, G. Canneto, F. Nanna, A. Villone, E. Fanelli, C. Freda, G. Cornacchia, G. Braccio, P.U. Foscolo, S. Heidenreich, E. Bocci; 1000 kWth gasification pilot plant with in-vessel high temperature gas filtration; 24th European Biomass Conference and Exhibition (EUBCE). Amsterdam, The Netherlands, 6–9 June 2016; <http://www.eubce.com/home.html>.

[72] Tuomi S, Kurkela E, Simell P, Reinikainen M. Behaviour of tars on the filter in high temperature filtration of biomass-based gasification gas. Fuel 2015;139:220–31.

[73] Stemmler M, Müller M. Chemical modeling of the "UNIQUE" process. In: Proceeding of the 1st international conference on polygeneration strategies, September 1–4. Vienna (Austria); 2009.

[74] Stemmler M, Tamburro A, Müller M. Laboratory investigations on chemical hot gas cleaning of inorganic trace elements for the "UNIQUE" process. Fuel 2013;108:31–6.

[75] Punjak WA, Shadman F. Aluminosilicate sorbents for control of alkali vapors during coal combustion and gasification. Energy Fuel 1988;2:702–8.

[76] Rieger M, Mönter D, Schulz RA. Removal of alkali from high temperature gases: sorbent reactivity and capacityDittler A.Hemmer G, Kasper G, editors. High temperature gas cleaning, vol. II. Karlsruhe: Institut für Mechanische Verfahrenstechnik und Mechanik Universität Karlsruhe (TH); 1999. p. 760–71.

[77] Dou BL, Shen WQ, Gao JS, Sha XZ. Adsorption of alkali metal vapor from high temperature coal-derived gas by solid sorbents. Fuel Process Technol 2003;82:51–60.

[78] Bachovchin DM, Alvin MA, DeZubay EA, Mulik PR. A study of alkali in pressurized gasification system. Process Report, Westinghouse Research and Development Center, Chemical Sciences Division, Pittsburgh (PA); 1986.

[79] Uberoi M, Punjak WA, Shadman F. The kinetics and mechanism of alkali removal from flue-gases by solid sorbents. Prog Energy Combust Sci 1990;16:205–11.

[80] Li YL, Li J, Jin YQ, Wu YQ, Gao JS. Study on alkali-metal vapor removal for high temperature cleaning of coal gas. Energy Fuel 2005;19:1606–10.

[81] Zheng Y, Jensen PA, Jensen AD. A kinetic study of gaseous potassium capture by coal minerals in a high temperature fixed-bed reactor. Fuel 2008;87:3304–12.

[82] Lee SHD, Johnson I. Removal of gaseous alkali–metal compounds from hot flue-gas by particulate sorbents. J Eng Power—Trans ASME 1980;102:397–402.

[83] Turn SQ, Kinoshita CM, Ishimura DM, Zhou J, Hiraki TT, Masutani SM. A review of sorbent materials for fixed bed alkali getter systems in biomass gasifier combined cycle power generation applications. J I Energy 1998;71:163–77.

[84] Turn SQ, Kinoshita CM, Ishimura DM, Hiraki TT, Zhou J, Masutani SM. An experimental investigation of alkali removal from biomass producer gas using a fixed bed of solid sorbent. Ind Eng Chem Res 2001;40:1960–7.

[85] Wolf KJ, Müller M, Hilpert K, Singheiser L. Alkali sorption in second generation pressurized fluidized-bed combustion. Energy Fuels 2004;18:1841–50.

[86] Diaz-Somoano M, Martinez-Tarazona MR. Retention of zinc compounds in solid sorbents during hot gas cleaning processes. Energy Fuels 2005;19:442–6.

[87] Rieß M, Müller M. High temperature sorption of arsenic in gasification atmosphere. Energy Fuels 2011;25:1438–43.

[88] Salmang H, Scholze H. Keramik Teil 1: Allgemeine grundlagen und wichtige eigenschaften, 1st ed. : Springer-Verlag; 1982.

[89] Barisano D, Freda C, Nanna F, Fanelli E, Villone A. Biomass gasification and in bed contaminants removal: performance of iron enriched olivine and bauxite in a process of steam/O_2 gasification. Bioresour Technol 2012;118:187–94.

[90] Jazbec M, Sendt K, Haynes BS. Kinetic and thermodynamic analysis of the fate of sulphur compounds in gasification products. Fuel 2004;83:2133–8.

[91] Westmoreland PR, Harrison DP. Evaluation of candidate solids for high-temperature desulfurization of low-Btu gases. Environ Sci Technol 1976;10:659–61.

[92] Yumura EM, Furimsky EE. Solid adsorbents for removal of hydrogen-sulfide from hot gas. Erdol Kohle Erdgas 1986;39:163–72.

[93] De Diego LF, Abad A, Garcia-Labiano F, Adanez J, Gayan P. Simultaneous calcination and sulphidation of calcium-based sorbents. Ind Eng Chem Res 2004;42:3261–9.

[94] Hu Y, Watanabe M, Aida C, Horio M. Capture of H_2S by limestone under calcination conditions in a high-pressure fluidized-bed reactor. Chem Eng Sci 2006;61:1854–63.

[95] Gasper-Galvin LL, Atimtay AT, Gupta RP. Zeolite-supported metal oxide sorbents for hot-gas desulfurization. Ind Eng Chem Res 1998;37:4157–66.

[96] Constant TK, Doraiswamy TL, Wheelock TT, Akiti TT. Development of an advanced calcium-based sorbent for desulfurizing hot coal gas. Adv Environ Res 2001;5:31–8.

[97] Álvarez-Rodríguez R, Clemente-Jul C. Hot gas desulphurization with dolomite sorbent in coal gasification. Fuel 2008;87:3513–21.

[98] Arvanitidis C, Zachariadis D, Vamvuka D. Flue gas desulfurization at high temperatures: a review. Environ Eng Sci 2004;21:525–47.

[99] Towler LG, Lynn LS, Fenouil LL. Removal of H_2S from coal-gas using limestone—kinetic considerations. Ind Eng Chem Res 1994;33:265–72.

[100] Lynn LS, Fenouil LL. Kinetic and structural studies of calcium-based sorbents for high-temperature coal-gas desulfurization. Fuel Sci Technol Int 1996;14:537–57.

[101] Aravind PV, de Jong W. Evaluation of high temperature gas cleaning options for biomass gasification product gas for solid oxide fuel cells. Prog Energy Combust Sci 2012;38:737–64.

[102] Li YL, Wu YQ, Gao JS. Study on a new type of HCl removal agent for high-temperature cleaning of coal gas. Ind Eng Chem Res 2004;43:1807–11.

[103] Verdone N, De Filippis P. Reaction kinetics of hydrogen chloride with sodium carbonate. Chem Eng Sci 2006;61:7487–96.

[104] Weinell CE, Jensen PI, Damjohansen K, Livbjerg H. Hydrogen-chloride reaction with lime and limestone—kinetics and sorption capacity. Ind Eng Chem Res 1992;31:164–71.

[105] Stemmler M, Tamburro A, Müller M. Chemical hot gas cleaning of syngas from biomass gasification for production of biofuels. In: Proceedings of ICPS10, international conference on polygeneration strategies, September 7–9. Leipzig (Germany); 2010.

[106] Duo W, Kirkby NF, Seville JPK, Kiel JHA, Bos A, DenUil H. Kinetics of HCl reactions with calcium and sodium sorbents for IGCC fuel gas cleaning. Chem Eng Sci 1996;51:2541–6.

[107] Nunokawa M, Kobayashi M, Akiho H. Halide compound removal from hot gasification fuel gas with sodium based sorbent. In: Proceedings of the GCHT-7, June 23–25. Shoal Bay (Australia); 2008.

[108] Antonini T, Gallucci K, Foscolo PU. A biomass gasifier including an ionic transport membrane system for oxygen transfer, International Conference on BioMass, 4–7 May 2014, Florence, Italy. Chem Eng Trans 2014;37:91–6.

[109] Antonini T, Gallucci K, Anzoletti V, Stendardo S, Foscolo PU. Oxygen transport by ionic membranes: correlation of permeation data and prediction of char burning in a membrane-assisted biomass gasification process. Chem Eng Process 2015;94:39–52.

Chapter 5

Advanced Process Combination Concepts

5.1 COMBINATION OF PYROLYSIS AND GASIFICATION

5.1.1 Multistage Gasification

Multistage gasification aims to perform pyrolysis and gasification separately and controlled. In this way the pyrolysis and gasification conversion steps can be improved by optimized operating conditions in each stage. Biomass comprises a high amount of volatiles. Typically, more than 80% of the biomass is volatile and the remaining part is charcoal [1]. Interaction between volatiles and char can have a negative impact on the reactivity and gasification of the char [2]. Thus, char gasification should be performed in the absence of the volatiles in order to increase the gasification efficiency with high carbon conversion rates. Modern, advanced gasification concepts separate the pyrolysis and the gasification steps in single controlled stages which are combined in a multistage gasification process. The different stages can be combined in one unit with separated pyrolysis and gasification zones which is realized by a staged supply of different amounts of gasifying agent at defined locations in a fixed bed or a fluidized bed gasifier. Alternatively, the different stages are realized by combining single reactors—a pyrolysis reactor and a gasifier combined in a series. Multistage gasification enables conversion of biomass under optimized operating conditions for the single steps. A syngas of high gas purity with low levels of tar can be achieved. Furthermore, the overall process efficiency with high char conversion rates and product quality and quantity can be enhanced by using multistage gasification compared to a one-stage gasification.

To produce a tar-free syngas, one of the first multistage gasifiers was developed by the Asian Institute of Technology in Thailand in 1994 [3]. The design was a two-stage downdraft gasification in a single reactor with two levels of air supplies, see Fig. 5.1. Compared to a one-stage reactor under similar operating conditions, the tar content was reduced by about 40 times from 3600 to 92 mg/m$_N^3$, the higher heating value was increased from 3.72 to 4.15 MJ/m$_N^3$ and the cold gas efficiency was increased from 60.8% to 69.2% [3]. A problem

Advanced Biomass Gasification. DOI: http://dx.doi.org/10.1016/B978-0-12-804296-0.00005-1

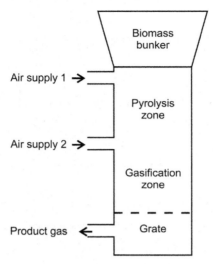

FIGURE 5.1   Principle scheme of the downdraft gasifier developed at the Asian Institute of Technology in Thailand in 1994 [3].

of the gasifier was that the whole reaction zone slowly moved down towards the grate and the height of the active reaction zone shrank. Some years later in 2001 Bhattacharya [4] reported on a hybrid biomass-charcoal gasifier with three stages of air supply which is based on a modified version of the multistage gasifier of Bui [3]. Bhattacharya achieved tar contents down to $28\,mg/m_N^3$ and a gasifier efficiency of 75.8% with this hybrid gasifier.

In recent years, several new and improved multistaged biomass gasifier concepts have been developed and studied (eg, Refs. [5–13]). The basic principle of these staged gasifiers is to produce char and volatiles which contain a high amount of tars in the first pyrolysis stage. The char is transported into the gasification stage and tar is reduced by partial oxidation (POX) or is steam reformed and additionally reduced by contact with char (by passing through a moving or fluidized char bed). Some recent studies have been performed at lab-scale and the feasibility of a scale-up of the gasifier to industrial scale is missing so far. However, some concepts have already been tested on a pilot scale or even on a larger scale. Examples which shall be described here in more detail are the Viking gasifier developed at the Danish Technical University, the FLETGAS process developed at the University of Sevilla in Spain, the LT-CFB (low temperature (LT) circulating fluid bed) gasification process developed by the Dong Energy company from Denmark, and the Carbo-V process developed by the Choren company in Germany.

The Viking gasifier (see Fig. 5.2) developed at the Danish Technical University [5] is a two-stage process with a screw pyrolysis reactor followed by a fixed bed downdraft gasification reactor. The outlet of the pyrolysis reactor is directly combined to the gasification reactor. Between the pyrolysis and the

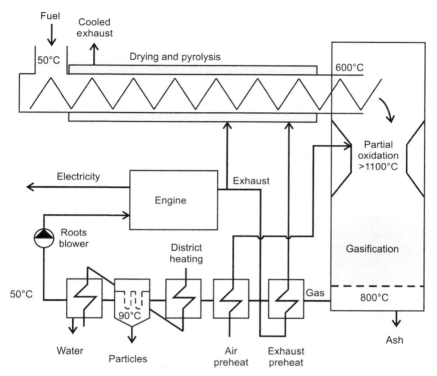

FIGURE 5.2   Principle scheme of the Viking gasifier [5].

gasification zone, air is added to partially oxidize the pyrolysis products. By the POX and by passing the char bed, the tar content in the syngas is reduced to less than $15\,mg/m_N^3$. The produced gas contains about 32% H_2, 16% CO, 20% CO_2, 30% N_2, and a small amount of CH_4 of about 2%. The higher heating value of the gas is about $6.6\,MJ/m_N^3$ [5]. A cold gas efficiency of 93% has been reported for the process [14]. The Viking gasifier has recently been scaled up to a $200\,kW_e$ gasifier and work is in progress for a $500\,kW_e$ gasifier [15].

A two-stage gasifier very similar to the Viking one has recently been studied in China by Wang et al. [16]. The influence of the oxygen concentration in the gasifying agent as well as of different biomass species on the gasification temperature, the gas composition, and the carbon conversion has been investigated. Cotton stalk pellets and pine sawdust pellets have been used as biomass feedstock. Carbon conversion efficiency of 80% has been achieved and the content of H_2 and CO in the syngas increased with the oxygen level in the gasifying agent. By increasing the oxygen level from 21 to 99.5 vol. %, the content of H_2 and CO in the syngas raised from 30 to 70 vol. %. The H_2/CO ratio was independent of the oxygen level in the gasifying agent and always about 1.

The FLETGAS process (see Fig. 5.3) developed at the University of Sevilla in Spain [12] is a special designed three-stage gasification process. In the first stage, devolatilization in a fluidized bed reactor takes place with low conversion of tar and char at temperatures between 700°C and 750°C. Air and steam can be added in this stage at reduced amounts to keep the temperature low to have just the devolatilization. A high amount of reactive tar is produced. In the second stage, the tar is reformed with steam at a high temperature (HT) of 1200°C. In the third stage, the char generated in the first stage is gasified in a moving bed downdraft reactor. The gas coming from the second stage flows through the char bed which serves as a catalyst for further tar reduction. The char produced in the first stage is directly transported from the first stage to the third stage via a gas-tight solid transport device [17,18]. Some modeling work [12] has been performed showing the advantages of the process compared to one-stage fluidized bed gasification. A significant decrease of tar concentration to $10\,\text{mg/m}_N^3$, char conversion of 98%, and gasification efficiency of 81% compared to tar concentration of $31\,\text{g/m}_N^3$, char conversion of 59%, and gasification efficiency of 67% for the one-stage fluidized bed gasification has been shown, which is an interesting result of this new concept. The composition of the gas was calculated on a dry basis as: 55% N_2, 13% CO, 15% CO_2, 4% CH_4, 8% H_2, and 2% C_2H_6. The higher heating value was calculated as $6.4\,\text{MJ/m}_N^3$. Some experimental work [19–21] has also been performed to develop the process. The process is still under development at the pilot scale [12]. Besides the proof of concept by

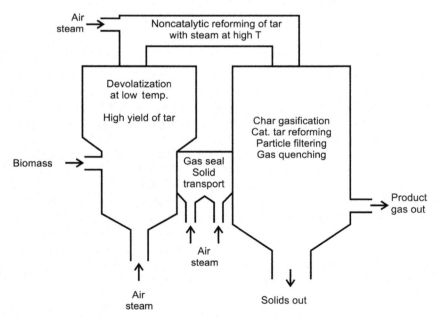

FIGURE 5.3   Principle scheme of the three-stage FLETGAS gasification process [12].

operation of a pilot unit a techno-economical assessment would also be needed to prove the economical benefit of this concept. A drawback seems to be the high complexity of the reactor set-up.

Fig. 5.4 shows the schematic principle of the LT-CFB gasifier developed by the Dong Energy company from Denmark [13]. The gasifier has two stages. The first stage is a circulating fluidized bed pyrolysis reactor operated at about 650°C. The second stage is a bubbling fluidized bed reactor operated at about 730°C for the gasification of char. Gasification of char is possible at this LT since the residence time in the gasifier is high. The gasifier is autothermally operated by using air as the oxidizing medium. Sand and ash is recirculated from the bottom of the gasifier to the pyrolysis reactor carrying the heat for the pyrolysis of biomass. Additionally, the char gas is redirected to the pyrolysis reactor. In between the two reactors, a cyclone is installed to separate char and sand from the gas. Char and sand enter the gasifier and pyrolysis and char gas are cleaned from ash in a second cyclone [13]. After testing the process in a $100\,kW_{th}$ [22] and a $500\,kW_{th}$ unit [14], a $6\,MW_{th}$ demonstration plant was built in 2012 at Asnaes power plant in Kalundborg in Denmark where the produced gas is cofired with coal [13]. Cold gas efficiencies of 87–93% have been achieved in tests with the $500\,kW_{th}$ unit [14]. A gas composition of 3.5% H_2, 16.3% CO, 14.5% CO_2, 59% N_2, 4.3% CH_4 and higher heating values of 5.2–$7\,MJ/m_N^3$ have been achieved [23]. It is reported that the two-stage process is robust and the construction is cheap and requires low maintenance. Alkalis are maintained in the ash due to the low process temperature [24]. However, the

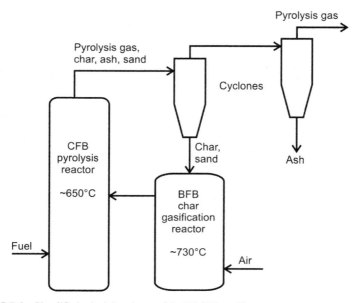

FIGURE 5.4   Simplified principle scheme of the LT-CFB gasifier.

produced gas has a high tar content ($>4.8 \, g/m_N^3$) [23]. Thus, use of the gas in engines, fuel cells, or for biofuel production is not possible without further gas cleaning.

The LT-CFB gasification process has been developed for the use of more difficult biomass feedstocks, such as straw, manure fibers, sewage sludge, and different organic wastes [14]. A comprehensive screening of low-grade biomass materials, such as vegetable residues (eg, bagasse, rice husks, olive kernel, palm kernel shells, empty fruit bunches), animal residues (eg, cattle and pig manure, meat and bone meal), waste water residues (eg, waste water sludge pellets, waste water fat fraction), and beach cleaning waste, for use in the LT-CFB gasification process has been performed by Thomsen et al. [25]. Bagasse, rice husks, olive kernel, olive wood prunings, and waste water sludge pellets showed the best overall potential and no apparent critical issues for use in the LT-CFB gasification process. Waste water fat fraction and waste water sand fraction can't provide sufficient energy for the pyrolysis due to their low carbon content. Palm kernel shells, meat and bone meals, beach cleaning waste, and vine wood prunings seem to be most problematic and unlikely to succeed.

There has been so far one multistage gasification process built and operated at a large scale which separates pyrolysis and gasification in different reactors— the Carbo-V process developed by the Choren company in Germany [26–28]. Fig. 5.5 shows a scheme of the three-stage Carbo-V process. The first process stage is a pyrolysis reactor, called the LT gasifier. The second process stage is a combustion chamber where the pyrolysis gas and recycled char from the deduster is oxidized with pure oxygen. The third process stage is a gasification reactor. Char produced in the pyrolysis reactor is gasified in this stage by using the combustion gas from stage 2 as the reacting agent. Since the reactions in the third stage are strongly endothermic, this stage is also called chemical

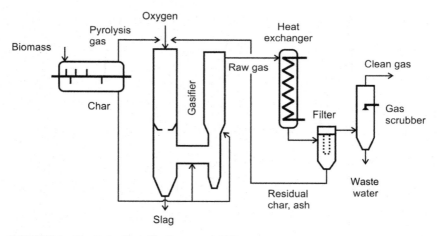

FIGURE 5.5 The Carbo-V gasification process [27].

quenching [27]. The process has been demonstrated in a $1\,MW_{th}$ plant in long-term operation for the production of biofuels. A cold gas efficiency of 82% and an almost tar-free gas have been reported for the process [28]. The gas composition was 34.6% H_2, 36.8% CO, 22.6% CO_2, 1.7% N_2, 0.4% CH_4, and 3.9% H_2O [29]. An upscale of the process to a $45\,MW_{th}$ plant was commissioned and a test program was started but not completed due to the insolvency of Choren in 2011 [30]. The main reasons for the stopping of the 45 MW plant have been seen in problems related to the scale-up of the new technology. This extended the commissioning of the plant far beyond the planned schedule and increased the costs so that the investors decided to stop their funding [31,32]. The technology has been acquired by the German company Linde in 2012 [33]. Linde reported, about 1 year after the acquisition of the Carbo-V technology, that they have refined the process to a level that demonstrations with a partner can commence [34]. In 2013, Forest BtL Oy from Finland, a subsidiary of the Finish Vapo Oy company, signed a license agreement with Linde for the Carbo-V process, which will be implemented in a new BtL (Biomass to Liquid) plant with a gasification capacity of 480 MW in Kemi, Northern Finland. Forest BtL Oy selected the Carbo-V technology since this multistage gasification process offers a high fuel conversion to syngas and a high syngas quality [35]. The plan was to start the BtL plant for the commercial production of 130,000 t/year of biodiesel and naphtha at the end of 2016. However, Vapo Oy froze the project for the Kemi plant in 2014 due to the uncertainty concerning the legislation on renewable fuels in the European Union (EU) and since the climate and energy strategy of the EU published in 2014 did not agree on new binding limits for the share of the renewable component in traffic fuels after 2020 [36].

The uncertain market situation for renewable fuels caused by legislation and combined with decreased costs for natural gas and crude oil due to the shale gas production in the United States was in general a major setback for biomass gasification in the EU and led to a stop for many large biomass gasification projects in 2014.

Table 5.1 gives an overview of the described multistage gasification processes. It can be concluded that very low tar concentrations in the producer gas are achieved by the different multistage gasification processes. Also, higher char conversion and gasification efficiencies are achieved compared to one-stage gasification processes. The complexity of the gasification process is increased by combining different reactors. However, this can be compensated by a simpler downstream gas cleaning process.

5.1.2   Use of Pyrolysis and Gasification at Different Locations

Biomass has a low bulk density resulting in high costs for transportation. Thus, it is preferred to use biomass available close to the gasification facility, which results in small decentralized plants. However, it is more economical to produce biofuels in large-scale plants. For this, biomass has to be transported over

TABLE 5.1 Overview of Multistage Gasification Processes

Gasification process	Number of stages	Process stages	Cold gas efficiency (%)	Tar content in the producer gas (mg/m$_N^3$)	Gas composition	Higher heating value (MJ/m$_N^3$)
Viking gasifier	2	• Screw conveyor pyrolysis reactor • Downdraft fixed bed gasifier • Partial oxidation by air addition between the 2 stages	93	<15	32% H$_2$ 16% CO 2% CH$_4$	6.6
FLETGAS process	3	• Fluidized bed pyrolysis reactor • Steam reformer • Moving bed downdraft gasifier	81	10	8% H$_2$ 13% CO 4% CH$_4$	6.4
LT-CFB gasifier	2	• CFB pyrolysis reactor • BFB gasifier	87–93	>4800	3.5% H$_2$ 16.3% CO 4.3% CH$_4$	5.2–7
Carbo-V process	3	• Pyrolysis reactor • Partial combustion chamber • Entrained flow gasifier	82	Tar free	34.6% H$_2$ 36.8% CO 0.4% CH$_4$	High

long distances which increases the transportation costs significantly and moreover has negative environmental impact. To avoid these conflicting drawbacks, a special concept to use pyrolysis and gasification at different locations has been developed under the name bioliq process by the Karlsruhe Institute of Technology (KIT) [37–39]. In this concept, pyrolysis aims to concentrate biomass at decentralized small pyrolysis plants for an economical transport of the biomass pyrolysis products (liquid and solid) to a centralized large gasification plant in order to produce biofuels (see Fig. 5.6). In the bioliq concept, biomass is concentrated into an oil–char slurry. The concentrated oil–char slurry is then transported to a large central process plant for gasification of the slurry and synthesis of biofuels. The concept is based on the use of low-grade lignocellulosic biomass, such as straw or forest residues. The energy density of the oil–char slurry is increased by about 10 times compared to the initial energy density of straw. Thus, transportation of the oil–char slurry is much more economical than that of the untreated biomass [40].

A demonstration plant of the bioliq process has been built at KIT in Germany in different construction stages from 2005 to 2013 [41]. The process comprises the following four process steps: (1) fast pyrolysis to produce the oil–char slurry in a separate decentralized prestep; (2) gasification of the slurry to produce syngas; (3) hot syngas cleaning and conditioning; and (4) synthesis of the biofuel.

Fast pyrolysis has been chosen since short reaction times give higher yields of pyrolysis oil [42]. The slurry is gasified in a $5\,MW_{th}$ Lurgi entrained flow gasifier at an operating pressure of up to 8 MPa [41]. Even if entrained flow gasifiers are widely used for gasification of water-based coal slurries or refinery residues, the gasification of such pyrolysis oil–char slurry is new and parts of several experimental investigations are related to the atomization of the oil–char slurry as well as to the modeling of the gasification of the slurry [43].

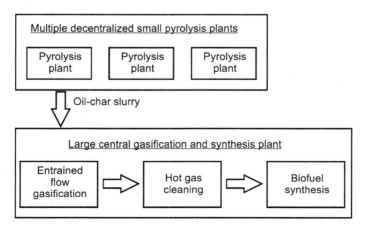

FIGURE 5.6 Schematic principle of the bioliq concept.

The syngas is cleaned in a completely dry hot gas cleaning process [44]. First, the syngas is filtered at 800°C in a special designed ceramic hot gas filter, where the filter elements are horizontally installed [45]. In a typical hot gas filter design, filter candles closed at one end are installed vertically hanging in a tubesheet [46]. An advantage of the design with horizontally installed filter elements is a more compact design of the filter with a smaller footprint and size of the filter vessel. A problem can be the regeneration of the filter elements which requires a very efficient backpulse system to remove the dust cake also from the top side of the horizontally installed filter elements. Downstream of the filter, chloride and sulfur gas components, such as HCl and H_2S, are removed by sorbents and tars are catalytically reformed in a following step.

One open issue which needs to be addressed by further research is how storage (storage conditions and duration) influences the properties and composition of the bio oil and accordingly the atomization and the gasification behavior of the oil slurry. Due to the composition, bio oil generally tends to change during its storage. Bio oil from pyrolysis of biomass is a mixture of different components, such as furfural, phenols, aldehydes, ketones, esters, ethers, etc. [47], having a quite high content of oxygen and water. Depending on the biomass feedstock, oxygen content is usually in the range of 35–40% and the water content is between 15% and 30% [48].

Furthermore, the properties of different oil–char slurries coming from different locations and different biomass feedstocks have to be tested and stable atomization and gasification have to be proven for these different slurries.

An economical assessment of the bioliq concept has shown that the production of gasoline and olefins from biomass by this process is not competitive compared to current market prices. Only subsidies, for example, from tax reduction and CO_2 certificates, could currently enhance the competitiveness of the biofuels [40].

5.2 COMBINATION OF GASIFICATION AND COMBUSTION

5.2.1 Dual Fluidized Bed Process with Internal Combustion

Biomass gasification is an endothermic process and requires providing thermal energy to the gasification reactor. This can be done in different ways that in most practical applications involve burning part of the gasification products, including sometimes auxiliary fuels made available by downstream purification treatments. Burning part of the gasification products is typically performed by using air, which is also used as a gasification agent. The ratio between biomass and air fed to the gasifier, commonly related to what would be needed by a stoichiometric biomass combustion process, is expressed as a fraction of it and called the *equivalence ratio*. In the case of fluidized bed gasifiers, the air flow rate should also comply with an adequate fluidization regime.

A major drawback of this simple and cost-effective arrangement is contamination with nitrogen of the product gas, with N_2 molar fraction values of 50% and more, resulting in a substantial reduction of the gaseous fuel heating value (lower heating value (LHV)) down to 4–$5\,MJ/Nm^3$, still acceptable for generating power with internal combustion engines, although unsuitable for more sophisticated and efficient energy applications (eg, fuel cells) and chemical syntheses. To avoid this problem, air could be replaced by oxygen (enriched air) and steam mixtures, with the additional benefit of enhancing tar reforming reactions and char gasification. However, this choice implies oxygen or enriched air availability. With small to medium size biomass gasification units operating at ambient pressure and readily available gas separation systems, air enrichment would be mainly accomplished by selective nitrogen–oxygen sorption systems requiring feed gas compression, that is compression of a volumetric gaseous stream up to five times larger than the oxidant used in the gasification process, with a substantial penalty of the whole energy efficiency. The use of oxygen plants would only be economic for large biomass gasification plants of several hundreds of Megawatts. For example, large commercial entrained flow coal gasifiers with a size of several hundreds of Megawatts are typically operated with pure oxygen.

An interesting approach to generate a product gas from biomass with a high heating value, is the dual fluidized bed (DFB) gasification process. The basic concept of this process is to combine two physically separated reaction zones— a gasification and a combustion zone in the fluidized bed system. The concept is schematically shown in Fig. 5.7. Additionally, heat and mass fluxes within the system are illustrated in the figure. Biomass is fed into the gasification zone and gasified with steam. Steam can easily be produced or is available even for small plants. Using steam as the gasifying agent increases the hydrogen content of the product gas and generates nitrogen-free gas with a heating value of typically between 10 and $16\,MJ/Nm^3$. Residual char and the mineral particle bed material (eg, sand, olivine) are transferred to the combustion zone where the char is burnt

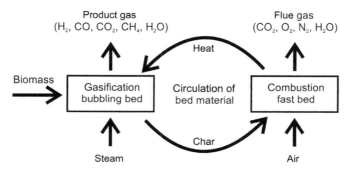

FIGURE 5.7   Basic concept of the dual fluidized bed gasification process.

with air. The heated bed material is recirculated from the combustion zone to the gasification zone transporting the heat required for the endothermic gasification reaction. The DFB process has attracted increasing attention from research as well as for industrial applications in recent years.

Even though the pioneering work of the DFB process was already started in the mid-1970s [49], the first industrial scale demonstration plants went into operation in 1999 and 2001, respectively. Several DFB designs have been developed worldwide [50,51] of which only two have been applied on an industrial scale for demonstration and commercial projects so far—the Güssing type DFB gasification process developed at the University of Vienna and the SilvaGas process initially developed by Battelle.

The well-known combined heat and power (CHP) plant in Güssing (Austria) was the first industrial demonstration plant of the DFB gasification process developed at the Institute of Chemical Engineering of Vienna University of Technology in the 1990s in cooperation with AE Energietechnik, and known internationally under the name FICFB (fast internally circulating fluidized bed) gasification system [52–55]. Other commercial plants where this process has been used so far are the 8.5 and $15\,MW_{th}$ plants in Oberwart and Villach (Austria), respectively, the $15\,MW_{th}$ plant in Senden/Ulm (Germany), and the $20\,MW_{th}$ plant in Gothenburg (Sweden).

The gasification reactor in the Güssing-type DFB gasification process is a bubbling bed fluidized by steam, where the biomass feedstock is devolatilized, and organic vapors and char are properly steam reformed to obtain permanent gases (CO, CO_2, H_2, CH_4). Fig. 5.8 shows schematically the design of the FICFB gasification reactor. The gasification temperature in the reactor bed is typically in the range between 850°C and 900°C. The granular bed material is continuously circulated by means of a chute and a loop seal to the combustion reactor which is operated as a fast circulating bed fluidized by air. Residual char transported along with the bed material is fully burnt in the combustion reactor, together with additional fuel properly injected, determining an increase in temperature of the particulate bed material of 50–100K above that in the gasifier. Bed material and flue gas are then separated at the top of the riser/ combustor in a cyclone-type device, and the hot bed material is brought back to the gasification chamber by means of a second loop seal that avoids gas mixing between gasifier and combustor, to provide the heat required by the gasification reactions. By adjusting properly the solid circulation rate, steady state behavior is obtained at the desired gasification temperature. As a result, air and steam can be used as gasification agents while keeping a very low nitrogen concentration in the fuel gas, permitting LHVs of about $12–14\,MJ/Nm^3$ and the utilization of steam as an additional gasification agent that improves gas quality—lower tar content and high hydrogen content of around 40 vol. %. A typical composition of the product gas for gasification of woody biomass is shown in Table 5.2 [54]. In comparison to more traditional installations, the complexity of plant layout and operation is increased, however the syngas quality is also increased

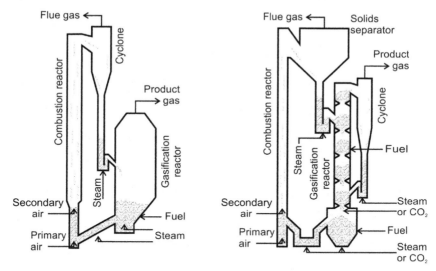

FIGURE 5.8   Design scheme of the conventional (left) and novel (right) FICFB gasification reactor developed by the TU Vienna.

TABLE 5.2 Typical Composition of the Product Gas for Dual Fluidized Bed Steam Gasification of Woody Biomass According to [56]

Component	Vol. %
H_2	40–43
CO	26–28
CO_2	19–21
CH_4	8.5–10
C_2H_4	1.5–2.5
C_2H_6	0.2–0.4
C_3H_8	0.1–0.2

remarkably, opening the way to more diversified utilization options and to polygeneration strategies.

The capacity of the Güssing plant is about 8 MW thermal input with electrical output of $2\,MW_e$ and district heating output of approximately $4.5\,MW_{th}$. A flow scheme of the Güssing plant including the fuel and flue gas cleaning

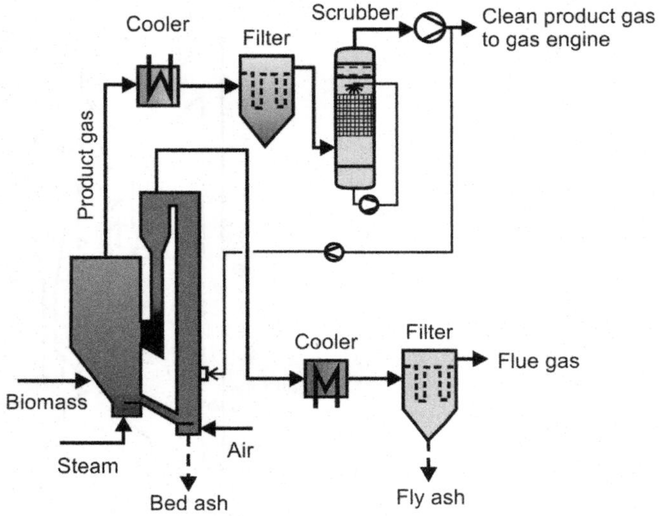

FIGURE 5.9 Flow scheme of the Güssing plant including the fuel and flue gas cleaning lines.

lines is shown in Fig. 5.9. The construction of the Güssing plant was started in September 2000, and electricity was first generated in April 2002. Since 2002, the plant has been operating regularly, and the present utilization index is above 7000 h/year. The gas cleaning set-up of the Güssing plant has been investigated and developed by RENET Austria (Energy from Biomass Network of Competence). Fine hydrated lime is injected into the product gas stream to obtain a primary reduction of tar to less than $1 \, g/m_N^3$, which allows enhanced recovery of sensible heat from the product gas stream. Solid particle removal is performed in the product gas filter which is operated smoothly at about 150°C. The remaining tar is removed from the product gas in a scrubbing column which is operated with biodiesel (RME, rapeseed oil methyl ester), to reach a tar content of about $20 \, mg/m_N^3$ and a temperature of about 50°C. Furthermore, water vapor contained in the product gas is condensed in the scrubber. To avoid disposal of liquid streams, the exhausted biodiesel is used as an additional fuel in the combustion reactor, and the condensed water is recycled to the gasification section. Heat from the flue gas of the combustion reactor is extracted by several heat exchangers. Fly ash is removed from the cooled flue gas by a fabric filter before releasing the gas into the atmosphere.

Very recently, an advanced design of the gasification section of the Güssing-type DFB reactor was proposed and tested at a pilot scale at the Technical University of Vienna [57]. It is characterized by intensive countercurrent contact between the bed material coming from the combustor and the volatiles coming from the fuel. This innovative design was developed especially for second generation low-grade biofuels and offers the possibility to reduce the tar content

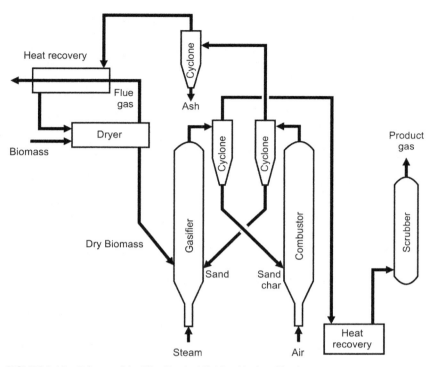

FIGURE 5.10   Scheme of the SilvaGas dual fluidized bed gasification process.

substantially in the gasification reactor. Fig. 5.8 shows schematically the major differences with the previous, conventional design.

The SilvaGas process comprises two circulating fluidized bed reactors—one is the gasifier and one is the combustor. The reactors are combined for heat transfer by circulating sand used as bed material between both reactors, see Fig. 5.10. This process has been initially developed by the Battelle Columbus Laboratories. In 1992, the process has been licensed to Future Energy Resources Corporation (FERCO). After successful testing of the process with over 20,000 h of operation in a 10 tons biomass per day research unit, a commercial scale demonstration plant of the SilvaGas process was built at the McNeil power station of Burlington Electric Department in Burlington, Vermont (United States) in 1997. This unit was designed for a capacity of 182 tons of dry wood equivalent to 42 MW thermal input [58]. After commissioning of the plant, the first run of the plant in full steam gasification operation was in August 1999. Continuous operation of the process was achieved in August 2000 [59]. Heating values of 17–19 MJ/m$_N^3$ have been achieved in extended operating periods. These heating values could be achieved independently of moisture changes of the used wood in the range from 10% to 50%. The increase of required heat with higher moisture content was provided by the circulating sand without any change of the gas composition. By supplying more heat for drying, the overall process efficiency decreased [59]. Even though the operation of the demonstration unit in Burlington was very successful and validated the performance of the SilvaGas process, no commercial unit of the SilvaGas process has been installed so far. The technology was acquired by Rentech Inc in 2009 and is still in their technology portfolio [60] even though they mothballed their activities in the alternative energy segment in 2013 [61].

Advanced developments of the DFB gasification technology aim to combine the gasification and the combustion zone in one reactor vessel by having the gasification zone placed concentrically surrounded by the combustion zone. One example of such a development is the small-scale high throughput biomass gasification system patented by FERCO [62] as a further development of the SilvaGas process. The system is schematically shown in Fig. 5.11. The idea and advantage of this system with concentric gasification and combustion zones is to increase the process efficiency by having direct heat transport from the combustion to the gasification reactor in addition to the heat transport by the circulating bed material as well as by reducing heat loss.

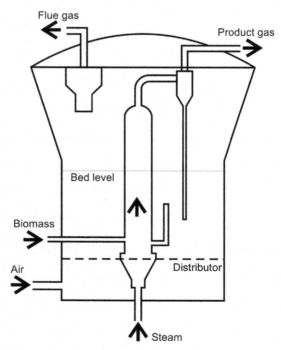

FIGURE 5.11   Scheme of the small-scale high throughput biomass gasification system of FERCO.

Another example of combining the gasification and combustion zones in one reactor vessel is the MILENA gasifier developed by ECN (Energy Research Centre of the Netherlands) in the Netherlands [63–65]. A scheme of the MILENA gasifier is shown in Fig. 5.12. The system has a compact design. The gasification zone comprises three parts—the riser, the settling chamber, and the downcomer. The combustion zone is operated as a bubbling fluidized bed. Bed material and unreacted char are transported from the gasification into the combustion zone by the downcomer. The bed material heated by combustion circulates through a hole at the bottom of the combustion zone into the riser. The MILENA design is relatively similar to the one of FERCO shown above in Fig. 5.11. Also in the MILENA design, the gasification zone is concentrically surrounded by the combustion zone. However, the system uses a settling chamber instead of a cyclone which is used in the FERCO design to remove bed material and unreacted char from the produced gas stream. The process achieves almost 100% carbon conversion and a cold gas efficiency of about 80%. The heating value of the produced gas is 12–15 MJ/m$_N^3$. ECN has built a 30 kW$_{th}$ MILENA lab-scale gasifier which was started in 2004. In 2008, ECN finished the construction of an 800 kW$_{th}$ MILENA pilot plant. ECN has licensed

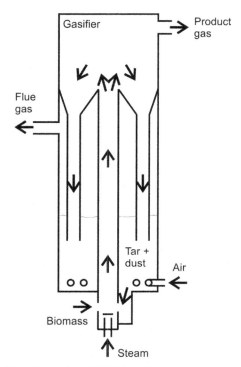

FIGURE 5.12   Principle scheme of the MILENA gasifier.

the MILENA technology to industrial partners. One of them, Royal Dahlman, is offering commercial systems and announced the signature for the development of an $18\,MW_{th}$ MILENA biomass gasification power plant in Shanghai (China) [66] and the construction of a $4\,MW_{th}$ plant in India [67]. Other commercial plants of the MILENA design are planned [68,69]. Before ECN built their first lab-scale $30\,kW$ gasifier, they started the development of the MILENA gasifier design by cold model testing [65].

Cold model testing is an easy approach to perform parameter studies and optimization as well as to support the design and construction of new prototypes for fluidized bed reactors. The dynamic behavior of a fluidized bed reactor is tested using a "smaller" cold model operated at ambient temperature and pressure. Similarity of the cold model and a gasifier reactor is given and expressed by a group of dimensionless numbers [70]. The concept of dynamic similarity as a design tool for a fluidized system has been validated in several experiments in the past [71]. Cold model testing has gained increasing interest from many researchers for the investigation of DFB biomass gasifiers in recent years (see eg, Refs. [72–75]), including the design innovations mentioned above in Fig. 5.10 [76].

Alternative DFB gasification designs with a reverse design compared to the FERCO and MILENA designs have been recently described and tested in a lab-scale unit by Miccio et al. [77] and by Simanjuntak and Zainal [78], respectively. The $20\,kW_{th}$ gasifier described by Miccio et al. [77] has the combustion zone concentrically surrounded by the gasification zone, see Fig. 5.13. The gasification zone is operated as bubbling fluidized bed and the combustion zone is a riser. The bed material and unreacted char is sucked from the gasification zone into the riser via orifices at the lower part of the riser tube. The bed material returns from the combustion zone to the gasifier via a second orifice on top of the gasification zone. Miccio et al. [77] reported on leakage problems between the gasification and the combustion zone in the test unit that have to be improved.

The gasifier described by Simanjuntak and Zainal [78] is very similar to the design described by Miccio et al. [77]. The differences are just the position and how the bed material is separated from the flue gas stream and recirculated to the gasification zone.

The bed material in a DFB gasifier can have different functions and a high influence on the quality of the produced gas. Beside its main function to transport heat from the combustion to the gasification zone, the bed material can additionally have reactive or sorption functions. It can enhance cracking of tar, prevent agglomeration of the bed or can remove gaseous impurities by sorption. The DFB gasification system is specifically well suited for sorption of gases by the bed material, since the bed material can easily be regenerated during its circulation between the gasification and the combustion zone. One example, where the sorption of a gas and the regeneration of the sorbent can advantageously be applied in the DFB gasification process, is the capture of CO_2. The use of a CO_2

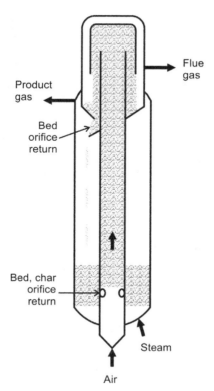

FIGURE 5.13   Example of a gasifier having the combustion zone concentrically surrounded by the gasification zone according to Miccio et al. [77].

sorbent minimizes carbon oxides and enhances the equilibrium conversion of catalytic tar and CH_4 steam reforming allowing to clean the gaseous fuel and to improve the H_2 yield. The advantages brought about by a reduction of the CO_2 content in the product gas by adsorption on basic oxides has been exploited in fundamental investigations of different bed materials comprising dolomite and nickel as sorbent and catalyst by Di Felice et al. [79]. The practical feasibility has been demonstrated by several investigations using the DFB gasification process and adding calcined limestone or dolomite to the reactor bed inventory [80–82]. The sorbent circulates between the bubbling fluidized bed of the gasifier—where CO_2 capture by carbonation takes place, and the riser of the combustor—where the sorbent is regenerated by calcination. The endothermic biomass gasification and the exothermic solid carbonation processes in the bubbling fluidized bed of the gasifier combine well together and their coupling reduces the amount of the solid circulation rate required to sustain thermally devolatilization and gasification reactions. On the other hand, the riser provides the calcined solid sorbent and the remaining thermal loading. When oxygen is utilized instead of air for the combustion reactions in the riser, a CO_2 stream can easily be obtained by steam condensation, available for storage and sequestration. Thus, a net removal of CO_2 from the atmosphere is achievable.

The thermodynamic constraints of the reaction between CO_2 and CaO impose, at ambient pressure, a temperature level for gasification somewhat lower (650–700°C) than the usual one (800–850°C). At the lower temperature of the gasifier bed, tar cracking by the bed material is less effective. However, the experimental evidence at pilot and industrial scale [80,81] does not show substantial increase of tar content in these conditions. The H_2 concentration in the dry product gas has been increased in pilot tests up to 75 vol. % [81]. In tests at industrial scale at the biomass gasification plant in Güssing, H_2 content of about 50 vol. % in the dry product gas has been achieved [82]. In recent modeling studies, H_2 content in the produced gas of up to 70% has been calculated [83]. The process of simultaneous biomass gasification and CO_2 capture by sorption is a promising approach for the production of hydrogen rich syngas. The process is often called in the literature "sorption-enhanced gasification" or "absorption enhanced reforming process (AER process)."

5.2.2 Gasification with Partial Oxidation

The production of tars in gasification processes is one of the major problems, since tars cause environmental hazards and create process-related problems like fouling, corrosion, erosion, and abrasion. Therefore, the tar content has to be reduced to values compatible with downstream processes. This can be done either by primary measures in the gasifier itself [84] or by secondary measures downstream of the gasifier. The latter comprise physical, thermal, and catalytic methods [85].

Thermal treatment of tar is one secondary method for the reduction of tar concentration in biomass gasification gas [85,86]. At HT, the stability of tar is reduced and tar can be converted or cracked into lighter gases. However, thermal reduction of tar is a quite complex process and the result depends highly on the process parameters. For example, it has been shown that just heating a tar-containing gas by an external heat source to a temperature of 900–1150°C leads to polymerization reaction of light tar components and to the formation of soot instead of the cracking of tars [87].

In recent years, partial oxidation (POX) of tars has been attracting the increasing interest of researchers as a method to achieve thermal tar conversion. In the case of POX, an oxidant is added to the producer gas and part of it is burned in a flame, so that mainly the tars, that is, CH- and CHO-compounds, will be partially oxidized to CO and H_2 preventing complete oxidation to CO_2 and H_2O. Air/fuel ratio, hydrogen concentration, methane concentration, temperature, and free radicals produced during the combustion influence the cracking or polymerization reactions of tar components. POX can be operated with and without a catalyst. Furthermore, steam can be added to the producer gas

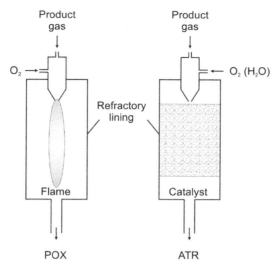

FIGURE 5.14   Schematic of POX and ATR reactor.

in the case of catalytic oxidation to enhance steam reforming reactions. In this case, the process is usually called autothermal reforming (ATR). In noncatalytic POX, reactions take place homogenously in the gas phase. Due to the uncatalyzed reactions, the gas temperature has to be higher in noncatalytic POX, which is achieved by higher oxygen addition, than in ATR, thus resulting in a decreased yield of synthesis gas. The concepts of POX and ATR are shown in Fig. 5.14 and the estimated gas compositions after the respective treatment are compared in Table 5.3 using the example of the Värnamo plant [88].

Houben et al. [87] investigated tar reduction by partial combustion using naphthalene as model tar component at a very low concentration of 2.6 mg/m_N^3 in a mixed gas stream of hydrogen, methane, and nitrogen. It has been shown that without oxygen addition polyaromatic hydrocarbons (PAH) and soot were formed. Also, if too much oxygen was added, PAH and soot were formed. For a very low air to fuel ratio of $\lambda = 0.2$, the partial combustion reduced the tar content by more than 90%. For $\lambda > 0.2$, the tar concentration increased. Thus, an optimum was found in the amount of air added for the reduction of naphthalene. For $\lambda > 0.4$, polymerization to higher tar rings and soot formation prevailed. It has also been found that hydrogen and methane have an influence on tar reduction. Hydrogen acts as an inhibitor for the formation of soot. With increasing hydrogen concentration, the amount of higher tar ring components decreased. At hydrogen concentrations higher than 20% nearly all naphthalene was converted to benzene or permanent gases. For higher methane concentrations and very low hydrogen concentrations, polymerization to higher tar ring components and soot formation was observed.

TABLE 5.3 Estimated Composition of the Producer Gas at VVBGC Plant at 15 bar [88]

Component	Gasifier (vol. %)	ATR at 1000°C (vol. %)	POX at 1300°C (vol. %)
Inlet O_2		7	10
Inlet temperature (°C)		800	800
C2-hydrocarbons	1.6	–	–
CH_4	8.2	–	–
CO	11.9	23.8	24.3
CO_2	27.9	19.8	19.2
H_2	11.8	23.0	16.1
H_2O	37.7	33.4	39.7
NH_3	0.3	0.2	0.2
H_2S	0.01	0.01	0.01
Tars	0.3	–	–
Mass flow ratio (kg/kg)	1	1.13	1.19
Supplied heat (MJ/kg)	–	0.25	0.37
LHV (MJ/m_N^3)	7.3	5.4	4.8
Efficiency		0.92	0.80

Naphthalene as a model tar component doesn't simulate the tar composition of real gases correctly. Real tar is composed of a lot of different components and has typically high concentrations of benzene and toluene which are the most stable tar species and require HTs above 1200°C for their effective reduction [89]. At the same time, HTs favor the formation of coke. Zhang et al. [89] showed that coke formation starts at 900°C having a maximum at 1100°C.

The influence of the hydrogen concentration on the presence of free radicals in the partial combustion process has been studied by van der Hoeven et al. [90]. Their investigation showed that a higher hydrogen concentration has a positive influence on tar cracking due to increased reaction rates, higher amounts of free radicals generated, and longer residence times of the radicals.

Svensson et al. modeled soot formation during POX of gasifier product gas [91]. They simulated different gas compositions and used naphthalene as a tar component. The simulation results showed a decrease of soot formation by increasing the hydrogen concentration in the gas as well as an increase of soot formation in the case of increased methane concentration. This is in qualitative agreement with the experimental results by Houben et al. [87]. However, the influence of the hydrogen concentration on the soot formation was very low in the simulation, which requires further investigations. In all simulated cases soot was formed and could not significantly be reduced except when no naphthalene was added to the gas. In this latter case about 50% less soot was formed compared to the base case where naphthalene was added in the amount of 0.5 mol. % of the total incoming gas flow.

Ahrenfeldt et al. [92] showed higher tar reduction with increasing air ratio for the POX of real gas from the pyrolysis of wood pellets. They showed complete conversion of phenol for $\lambda >0.3$ and temperatures above 950°C. For lower air ratios higher temperatures were required to get the same result. Above 900°C phenol and other primary tars were converted into low molecular weight PAH, primarily naphthalene. The tar concentration in the raw pyrolysis gas was quite high (above 200 g/kg wood).

Investigations by Brandt et al. [93] showed a minimum tar concentration at an air ratio of $\lambda = 0.5$ for POX of gas from the pyrolysis of straw. For higher air ratios the tar concentration slightly increased. The tests were performed at 800°C and 900°C. The tar reduction was nearly comparable for both temperatures. Composition of tar downstream of the POX reactor as well as soot formation has not been investigated.

Su et al. [94] investigated the POX of gas from the pyrolysis of rice straw. Maximum tar reduction has been achieved at an air ratio of $\lambda = 0.34$ at a temperature of 900°C. Phenolic compounds were completely converted for $\lambda >0.2$. It is reported that oxygen increases the reaction rate and can promote the formation of free radicals for initial tar cracking reactions. For air ratios above 0.278 only PAHs, that is, naphthalene, fluorene, phenanthrene, and pyrene, can be found in the gas. Higher air ratios lead to formation of higher PAHs. This agrees with the results of Houben [87].

While Houben [87] used a microswirl burner in his investigations, Weston et al. [95] determined the suitability and effectiveness of a Coandă ejector, which makes use of the Coandă effect, the phenomenon in which a jet passed over a curved surface will attach to the wall. In a gas produced by pyrolysis of wood pellets at 800°C, benzene was reduced by 95%, toluene was reduced by 96%, naphthalene was reduced by 97.7%, and the gravimetric tar yield was reduced by 86.7%. The average temperature recorded in the flame zone of the Coandă tar cracking system was slightly below 1000°C and the highest temperature recorded was 1290°C. The temperatures recorded are high enough for thermal cracking of tar to take place.

POX for tar reduction has already been applied successfully in some recent multistage gasification concepts, such as the Viking gasifier (see Section 5.1) for which a tar reduction by partial combustion by a factor of 100 has been reported [5] or the newly developed Fraunhofer ISE gasification process for production of synthesis gas with tar contents below 50 mg/m_N^3 [96]. The partial combustion in these multistage gasifiers aims at increasing the temperature of the pyrolysis gas required for the subsequent endothermic char gasification reaction.

In a single stage gasification process, a noncatalytic POX stage downstream of the gasifier is typically not used since the HT required for an efficient conversion of tars reduces the energy efficiency of the process. Additionally, it reduces the syngas yield. However, catalytic POX as a catalytically active thermochemical process can provide an effective way for tar removal and conversion. There are also problems associated with catalytic POX, such as poisoning of catalysts by gas impurities, for example, H_2S, or coke formation.

Different catalysts for (partial) oxidation of tars derived from conversion of biomass have been investigated in recent years. Carnö et al. investigated the oxidation of hydrocarbons from wood combustion with Pt and CuOx/MnOx supported on alumina [97]. The Pt-catalyst, with 85–95% conversion at >400°C, had a higher activity for naphthalene than the metal oxide catalyst, and deactivation was also less pronounced. In a comparative investigation by Miyazawa et al. on catalytic POX of tar from biomass pyrolysis, $Rh/CeO_2/SiO_2$ and Ni catalysts exhibited much higher performance than other materials, such as active clay, USY zeolite, MS-13X, dolomite, alumina, silica sand, and fluorite [98]. However, $Rh/CeO_2/SiO_2$ exhibited higher and more stable activity under the presence of high concentration (280 ppm) H_2S than the Ni catalyst [99]. Ni/CeO_2 showed a smaller amount of coke and thus less deactivation than Ni/Al_2O_3, Ni/ZrO_2, Ni/TiO_2, and Ni/MgO [100]. Wang et al. found 99% conversion of naphthalene using a monolithic Ni-catalyst [101]. Herrmann found complete conversion of naphthalene above 500°C using a $Mo_8V_2W_1O_x$-catalyst [102]. Ammendola et al. found high catalytic activity of Al_2O_3 supported Rh, $LaCoO_3$, and $Rh-LaCoO_3$ catalysts for conversion of tars from maple wood pyrolysis [103]. Böhning found almost complete conversion of tar from fixed bed wood gasification above 700°C using Pd- and Ni-based catalysts [104].

In small- and medium-scale gasification systems, which are usually relevant for biomass gasification, the supply of oxygen for POX is very often realized by simply injecting air rather than pure oxygen into the producer gas due to cost reasons. However, the producer gas is significantly diluted when using air, resulting in a low heating value which is by way of an example insufficient for proper use in a gas engine [104]. Therefore, a combination of an oxygen carrier material (OCM) with a catalyst for POX has been proposed by Ma et al. [105]. In this method, called chemical looping partial oxidation (CLPOX), the amount of oxygen needed for the POX of tar will be supplied by an OCM, which will also serve as a support of the catalyst. OCMs, well known from application in chemical looping combustion [106], are capable of taking up oxygen at high oxygen partial pressure in air and releasing oxygen at low oxygen partial pressure in the producer gas, thus providing oxygen without dilution by nitrogen. In the aforementioned investigations [105], mainly perovskite-type materials were used as OCM due to their relatively high oxygen storage capacity. In laboratory tests, the perovskite material CSFM5555 ($Ca_{0.5}Sr_{0.5}Fe_{0.5}Mn_{0.5}O_{3-\delta}$) and ilmenite exhibited promising properties as oxygen carriers. They can supply a sufficient amount of oxygen for the oxidation of tar and show excellent redox properties for the regeneration. Among different investigated catalyst materials, that is, several perovskite-type materials and one Ni-based catalyst, which should be used in combination with an OCM, NiO showed the highest catalytic activity in POX of naphthalene as a model tar compound [107]. It showed high tolerance against pollutants usually present in producer gas from biomass gasification, such as HCl and KCl, but was poisoned by H_2S, which could be prevented by absorption of H_2S by a CuO-based sorbent. In addition to H_2S reduction, the sorbent increased the catalytic activity of the entire system significantly, so that naphthalene was almost completely converted at temperatures above 450°C despite the presence of 140 ppm H_2S [107]. The combined system consisting of these three materials (ie, CSFM5555, NiO, and CuO-based sorbent) showed high ability for the removal of naphthalene. A naphthalene conversion ratio of more than 99% was achieved at 600°C for a reducing gas loaded with a naphthalene concentration of about 2 g/m³ [105]. The CLPOX system (Fig. 5.15) shall consist of at least two reactors, one operating under POX mode and the other(s) being regenerated with air at the same time [108]. During regeneration not only the OCM will be oxidized but also coke formed on the catalyst will be burned and the Cu-based sorbent will be regenerated by oxidation of the formed copper sulfide and subsequent decomposition. Preliminary tests of the system

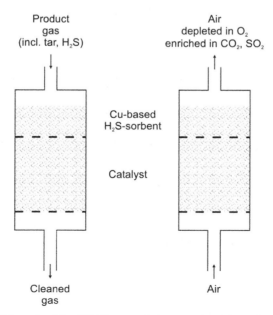

FIGURE 5.15   Schematic of the CLPOX reactor during operation (left) and regeneration (right).

in combination with a bench-scale fixed bed gasifier [108] show that the content of CO and H_2 in the producer gas was increased due to the POX of tars. The tar content was reduced by up to 98.5% and the heating value slightly increased by 2%. Total oxidation of the producer gas seems not to be an issue.

One drawback of CLPOX is the need for regeneration of the OCM, making the process more complex. Therefore, as an advancement of the process, the application of ceramic oxygen transport membranes (OTM), that is, mixed ionic and electronic conducting (MIEC), in combination with POX catalysts is under investigation. MIEC membranes work at temperatures above 700°C. Oxygen transport is driven by the difference in oxygen partial pressure across the membrane [109]. Oxygen adsorbs at the surface and decomposes into ions, which are transported through the membrane by a vacancy transport mechanism that is counterbalanced by the simultaneous flow of electrons in the opposite direction [110,111]. In a respective membrane reactor, the catalyst is in direct contact with the membrane or even coated on the surface of the membrane. The necessary oxygen for POX is supplied via the membrane and its amount can be controlled by the air fed into the membrane reactor. The utilization of ceramic OTMs to transfer the required oxygen input is made possible by the recent, substantial improvements in the membrane preparation methodologies [112] and in the oxygen permeation fluxes in the temperature range comprising that of catalytic POX [113].

5.2.3 Implementation of Gasification in Coal-Fired Power Plants

To substitute fossil fuels by renewable ones, biomass is cofired in coal-fired power plants [114–116]. This is a simple way for the reduction of fossil CO_2 emissions. One option which seems to be at the first sight the easiest and least expensive way is direct cofiring of biomass in the boilers. Up to about 3% of biomass on an energy basis can be directly cofired with minimal additional investment costs [117]. Direct cofiring of higher amounts of biomass can result in several problems. Biomass has quite different properties compared to coal, for example, biomass has lower ash content, higher oxygen content, higher content of volatiles, lower density and heating value. Moreover, grinding of biomass is more difficult due to fiber structures compared to grinding of coal and some kinds of biomass have high alkali and/or halogen content [118]. Problems which occur for direct cofiring of biomass in coal-fired boilers are mainly related to the different properties of the biomass ashes due to high alkaline and chlorine contents. Corrosion, slagging, and fouling in the boiler as well as in the heat exchanger and in the piping, poisoning of DeNOx catalysts and performance problems in electrostatic precipitators are the main drawbacks that are reported [119–122].

One approach to overcome these problems is indirect cofiring of biomass by biomass gasification. Biomass is gasified in a gasifier and the produced gas is cofired in a coal-fired boiler [123]. Up to 10% of the thermal capacity of the boiler can be cofired without the need of reconstruction of the boiler and auxiliary devices. For higher ratios major changes of the boiler would be required [124].

The first power plant where this concept was used is the Kymijärvi power plant in Lahti (Finland) [124,125]. The power plant has an electrical power production capacity of 167 MW and a district heating production capacity of 240 MW. A Foster Wheeler CFB (circulating fluidized bed) gasifier was installed for the gasification of biomass. Start-up of the gasifier was at the beginning of 1998. Depending on the feedstock and its moisture content the capacity of the gasifier is between 40 and 70 MW_{th} (nominal capacity 50 MW_{th}). Raskin et al. [125] reported on the successful operation of the plant. Wood chips, wet and dry waste wood, saw dust, bark, shredded tires, plastics, as well as recycled fuels were successfully gasified. The recycled fuels comprised mainly paper, cardboard, and wood, and about 5–15% plastics. It was demonstrated that the concept of indirect cofiring of biomass by gasification is a successful way to generate heat and power by using different types of biomass and a way to operate an existing coal-fired boiler with only small modifications. The power plant could be operated in a flexible way by adjusting the amount and the type of biomass which was cofired.

Based on the successful operation in Lahti, the concept of indirect cofiring of biomass by gasification has also been applied at the coal-fired Ruien power plant of Electrabel in Belgium by installing a CFB Foster Wheeler gasifier similar to the one in Lahti [126–129]. The gasifier was commissioned in December 2002 and commercial operation started in May 2003. The gasifier was installed at the boiler of unit 5 of the power plant. Unit 5 has a power output of 190 MW_e on coal. No pretreatment (drying, grinding) of the biomass as well as no cleaning and conditioning of the produced gas was required. The quality of the gas is sufficient to be burned in two burners, which are installed below the coal burners in the boiler. Wood residues from the local wood industry as well as fresh wood chips were used as feedstock. About 9% of the coal was substituted by gas from the biomass gasification and 120,000 t/year CO_2 emissions were saved [129].

An 85 MW_{th} CFB Lurgi gasifier has been in operation for indirect cofiring of biomass at unit 9 (net production capacity: 600 MW_e and 350 MW_{th}) of the Amer power station of Essent in Geertruidenberg in the Netherlands since 2002 [130,131]. Low quality demolition wood is mainly used as feedstock. About 90,000 t biomass was gasified in 2010 generating an electric capacity of about 33 MW_e [131].

A 10 MW_{th} CFB biomass gasifier designed by Austrian Energy was commissioned for indirect cofiring of biomass with coal at the Zeltweg power plant (137 MW_e, 344 MW_{th}) in Austria in 1997 [132,133]. Bark and wood chips were mainly gasified. The gasifier was shut down in April 2001 due to the closure of the power plant.

The interest in the concept of indirect cofiring of biomass by gasification declined after 2003. However, the interest in this concept has increased recently again and new biomass gasifiers were installed at coal-fired power plants. In Denmark, for example, Dong Energy installed a $6\,MW_{th}$ LT-CFB biomass gasifier (see Section 5.1), which has been in operation since 2011, at its Asnaes power plant [134]. Mainly straw is used as the biomass feedstock at this plant. The world's largest biomass gasifier so far was inaugurated in March 2013 in Finland [135]. The gasifier, a $140\,MW_{th}$ CFB gasifier designed and supplied by Metso (now Valmet), was installed at the $560\,MW_{th}$ coal-fired power plant of Vaskiluodon Voima Oy in Vaasa. Mainly harvesting residues and stumps are used as the biomass feedstock. Locally available peat is used as a back-up fuel. Biomass gas replaces between 25% and 50% of the coal consumption of the plant depending on the boiler load. A reduction in CO_2 emissions of about 230,000 t/year is achieved by using the biomass gasifier [136].

Indirect cofiring of biomass by gasification is a concept which is well proven by long-term operating experience in large coal-fired power plants. The advantages of this concept can be summarized as follows:

- Easy and cost-effective way to reduce fossil CO_2 of coal-fired power plants.
- Purity and quality of the produced gas can be low to be burned in the boiler. Thus, no high efficient gas cleaning and conditioning or cooling equipment is required and the investment costs are lower than in cases where high gas quality and purity are needed.
- High flexibility in using a broad range of biomass feedstocks, including low quality biomass as well as refuse-derived fuels.
- Biomass which generates problems in direct cofiring, for example, straw, can be used.
- Compared to direct cofiring, no drying or grinding is required.
- High biomass conversion; residual fine char particles and tars are burnt in the boiler.
- No operating dependency on availability of biomass; biomass can be substituted by fossil fuel.
- No influence on the operating availability of the power plant by possible problems with the gasifier.
- Less coal ash pollution than in direct cofiring (coal ash is often used in concrete production). Less biomass ash enters the boiler since coarser ash particles are removed in the cyclone of the CFB gasifier and returned into the gasifier where they accumulate and are removed as bottom ash. If higher percentages of biomass are cofired or higher coal ash purity is needed, a hot gas filter [46] can additionally be used to remove fine ash particles from the produced gas without energy loss due to cooling.

5.3 COMBINATION OF WATER GAS SHIFT CATALYSTS AND H_2-MEMBRANES

The production of synthetic fuels from producer gas requires a certain H_2/CO-ratio in the synthesis gas. By way of example, a ratio ≥ 3 is needed in case of methane production (compare reverse reaction 2.7). Usually, the ratios obtained in producer gases are far lower and need to be adjusted after the gasification process. To increase the amount of H_2 in a syngas or even to produce pure H_2 from a producer gas, the water gas shift (WGS) reaction is a well-known, industrially used process. By this reaction, water reacts with CO to form H_2 and CO_2:

$$CO + H_2O \leftrightarrow CO_2 + H_2 \quad \Delta_r H^0_{298} = -41\,kJ/mol \quad (5.1)$$

While this reaction is not influenced by pressure, since there is no difference in the number of moles between reactants and products, it is thermodynamically favored at lower temperatures due to its slightly exothermic character. However, the reaction is kinetically limited at LTs. Only at temperatures above 950°C is the reaction rate reasonably fast enough to operate without a catalyst [137]. Therefore, the reaction is usually performed catalytically at temperatures between 200°C and 450°C. To maximize the conversion rate, in industry the WGS reaction is performed with a high excess of steam in two stages with intermediate cooling. The first reactor is characterized by HT, 400–450°C, using Fe–Cr oxides as a catalyst, the second reactor operates at LT, 200–250°C, using Cu–Zn oxides as a catalyst. Thus, the CO-content in the gas stream can be reduced to below 4 vol. %. The output stream from the second reactor is directed to some hydrogen separation unit. Separation of CO_2 and H_2 is usually performed by physical or chemical scrubbing of CO_2 at LT and high pressure. Regeneration of the solvent is done at higher temperature and lower pressure. A further method is pressure swing adsorption using by way of example zeolites, activated carbon, or metal–organic compounds [138]. However, the necessary temperature and pressure changes cause efficiency losses of about 10%.

An alternative to the WGS reaction performed in conventional reactors is the utilization of membrane reactors combining the chemical reaction and the separation in a single unit (Fig. 5.16). Thus, the reaction can be performed in a single step even above the temperatures of the LT-shift of 250°C. By continuously removing one reaction product, for example, H_2, from the gas, the equilibrium of the WGS-reaction (5.1) is shifted to the product side according to the principle of Le Chatelier. Thus, high conversion rates of CO and H_2O to CO_2 and H_2 can also be achieved at HTs without the need for a high excess of steam. The driving force for H_2-permeation (or CO_2-permeation) being a difference in H_2 (or CO_2) partial pressure between the feed side and permeate side of the membrane can be achieved either by using a sweep gas, by a difference in absolute pressure between both sides, for example, high pressure on the feed side or

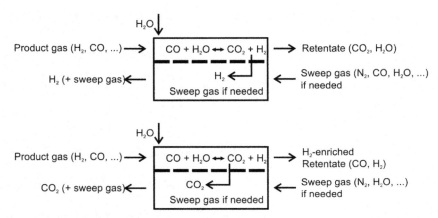

FIGURE 5.16 Schematic of a WGS-membrane reactor with H_2-permeable (top) and CO_2-permeable (bottom) membrane.

vacuum on the permeate side, or by a combination of both. The choice depends on the further use of the permeated H_2 (or residual syngas).

To separate hydrogen and carbon dioxide, the permeability of the membrane for the two gases must be as different as possible. Some work has been done on the development of CO_2-selective membranes due to the advantage of having still compressed hydrogen after gas separation [139]. Thus, compression of hydrogen, which is required for several chemical processes or combustion in a gas turbine, is not necessary. Furthermore, it is possible to get a synthesis gas as retentate consisting of CO and H_2 that has a suitable H_2/CO-ratio for a certain synthesis process. Usually, polymer membranes are used [140], which have a higher solubility for CO_2 molecules than for other, nonpolar gas molecules, such as N_2, H_2, and CO [141]. The diffusion rate of hydrogen is higher than that of carbon dioxide. Therefore, the material properties need to be adjusted to achieve mainly CO_2 permeation. As long as a partial pressure gradient of CO_2 over the membrane exists, CO_2 is continuously transported through the membrane. The main problem of CO_2-selective polymer membranes is their relatively low operation temperature due to decreasing selectivity and stability of the materials with increasing temperature [139]. The problem of decreasing CO_2-selectivity with increasing temperature applies also to microporous inorganic membranes which are more stable at higher temperatures than organic membranes [142,143]. Therefore, a direct combination of WGS reaction and CO_2-membrane using the aforementioned membranes in one single reactor cannot be easily realized.

Consequently, HT CO_2-membranes need to be developed. Recently, Wade et al. used mixed conducting materials made from molten carbonate salts immobilized within porous solid conducting oxides for separation of CO_2 from hot process gases at temperatures above 500°C [144]. In this type of membrane,

CO_2 from the gas first combines with an oxide ion from the solid oxide phase and is then transported as a carbonate ion through the molten carbonate phase. At the permeate surface, the carbonate ion releases CO_2 and the oxide ion returns to the solid oxide electrolyte, so that it can travel back to the feed surface of the membrane. These composite materials achieved extremely high CO_2 selectivity and good CO_2 permeability.

Much more work has been done to develop H_2-selective membranes. They can be divided into four types of membranes according to the used material and transport mechanism: polymeric, dense metallic, microporous ceramic, and proton-conducting dense ceramic membranes. Table 5.4 gives an overview of several types of H_2-membranes and their most important properties.

As already discussed for CO_2-selective membranes, gases are dissolved in polymeric membranes and transported through by diffusion. While in the case of CO_2-selective membranes the good solubility of CO_2 in the membrane material is utilized and the diffusion is under optimization, in the case of H_2-selective membranes the high diffusibility of H_2 is utilized and the low solubility needs to be increased. Again, selectivity and operation temperature are the limiting

TABLE 5.4 Overview on Different Types of Hydrogen Membranes [145–147]

	Polymeric	Dense metallic	Microporous ceramic	Proton-conducting dense ceramic
Materials	Organic polymers	Palladium alloys	Silica, alumina, zirconia, titania, zeolites	Perovskites
Transport mechanism	Solution-diffusion	Solution-diffusion	Molecular sieving	Solution-diffusion
Temperature (°C)	<100	300–700	200–600	600–900
H_2 selectivity	Low	>1000	5–139	>1000
H_2 flux (10^{-3} mol/m^{2}s at 1 bar)	Low	60–300	60–300	6–80
Stability issues	Swelling, compaction, mechanical strength	Phase transition (causes embrittlement)	Stability in H_2O	Stability in CO_2
Poisoning	HCl, SO$_x$, CO	H_2S, HCl, CO		H_2S

factors. Several materials have been tested in recent years. By way of example, a H_2/CO_2-selectivity of 100 was achieved by Chung et al. using modified 6FDA-Durol [148]. Pesiri et al. achieved a H_2/CO_2-selectivity of 20 using a polybenzimidazole membrane at 270°C, which would be a sufficient temperature for combination with LT-WGS catalysts [149].

In case of dense metallic membranes, hydrogen adsorbed at the surface dissociates into two hydrogen atoms which can diffuse through the metal lattice. Since other gas molecules usually do not dissociate at the metal surface, the selectivity of dense metallic membranes is very high [150]. For a long time, much work has been done on the development of membranes based on palladium or palladium-based alloys, for example, Refs. [151–156]. Furthermore, the application of other metals, such as copper [150], nickel [157], iron, platinum [158], and vanadium–nickel alloys [159] was investigated. The major problems of metallic membrane materials are hydrogen embrittlement [160] and poisoning by H_2S [161].

Microporous inorganic membranes for H_2-separation have a controlled pore size that is adjusted in such a way that the small molecule H_2 can diffuse through the material whereas bigger molecules like CO_2 cannot pass. Several materials were investigated, that is, silica [162], zeolites [163], and carbon [164]. The advantages of these materials are their relatively low cost and good stability under dry conditions. The disadvantages of these materials are their low selectivity and their instability in the presence of moisture, which is necessary for the WGS reaction [162].

Perovskites are often used as material for dense ceramic membranes. High purity H_2 can be achieved due to a proton transport mechanism. However, they have to operate at temperatures above 600°C. The general formula of perovskite-type oxides is $AB_{1-x}M_xO_{3-\delta}$. Base materials are $SrCeO_3$ [165,166], $BaCeO_3$ [167,168], or $BaZrO_3$ [169]. The B-element is partly substituted by trivalent oxides like Y, Yb, Nd, Gd, or La to produce oxygen vacancies. These enable the transport of hydrogen as hydroxide ions [170]. Other promising materials are La_6WO_{12} [171] and $LaNbO_4$ [172] having fluorite structure.

The combined catalyst-membrane system needs to be carefully selected since its overall performance does not only depend on the individual performance of the single components but also on the interactions between them under their respective operation conditions. Optimum conditions for one component might be different from the optimum conditions for the other component. Thus, the best compromise regarding the components and operation conditions to achieve the best possible overall performance needs to be identified. Furthermore, a suitable reactor design needs to be chosen. An overview on several membrane reactors can be found in [145]. Some examples for recent experimental investigations on combined catalyst–membrane systems utilizing mainly Pd-based membranes are discussed below.

Giessler et al. investigated in lab-scale experiments the performance of a microporous silica membrane packed bed reactor using a $Cu/ZnO/Al_2O_3$

catalyst under the conditions of the LT-WGS reaction [173]. The best permeation results were H_2 permeances of 1.5×10^{-6} mol/s/m^2/Pa, H_2/CO_2 selectivities of 8, and H_2/N_2 selectivities of 18. They reached 99% CO conversion during operation with a sweep gas flow of 80 cm^3/min, a feed flow rate of 50 cm^3/min, and a H_2O/CO molar ratio of 1 at 280°C. This is well above the thermodynamic equilibrium. Hydrophilic membranes underwent pore widening during the reaction while hydrophobic membranes did not suffer such behavior and also showed increased H_2 permeation with temperature.

Forster and van Holt et al. investigated mixed proton and electron conducting dense ceramic materials, that is, several barium cerates and zirconates and lanthanum tungstate, and iron–chromium oxide-based catalysts for application in a HT-WGS reactor ($\geq$600°C) [174–176]. Among perovskites, $BaCe_{0.2}Zr_{0.7}Yb_{0.08}Ni_{0.02}O_{3-\delta}$ showed a very promising stability at 600–900°C in a synthetic producer gas consisting of 15% H_2, 34% CO, and 51% H_2O. $La_{5.5}WO_{12-\delta}$ showed high stability at 900°C. Both materials showed also high stability in the presence of several pollutants like HCl, H_2S, and alkali compounds. Despite the HTs, a catalyst was needed to achieve a sufficient reaction rate. A Fe–Cr oxide-based catalyst showed high activity and high stability even in contact with the aforementioned poisons at temperatures above 600°C. Direct application of the catalyst on the surface of the membrane caused significant reactions between the two materials. However, in a membrane reactor with the catalyst as packed-bed no interactions were observed.

Augustine et al. tested a tubular WGS catalytic membrane reactor consisting of a porous Inconel supported, electroless plated Pd-membrane packed with 48–60 mesh iron–chrome-based catalyst [177]. Experiments were conducted with a CO and steam feed at 350–450°C, 1.44 MPa ($P_{tube} = 0.1$ MPa), H_2O/CO ratios of 1.1–2.6, and GHSVs (Gas Hourly Space Velocities) of up to 2900 h^{-1}, considering the effect of the H_2O/CO ratio as well as temperature on the reactor performance. Further experiments were conducted with a simulated syngas feed at 1.4 MPa ($P_{tube} = 0.1$ MPa), and 400–450°C, assessing the effect of the space velocity on the reactor performance. A maximum CO conversion of 98.2% was achieved with an H_2 recovery of 81.2% at 450°C. An optimal operating temperature for high CO conversion was identified at approximately 450°C, and high CO conversion and H_2 recovery were achieved at 450°C with high throughput, made possible by the 1.44 MPa reaction pressure.

Bi et al. tested a $Pt/Ce_{0.6}Zr_{0.4}O_2$ catalyst in a reactor furnished with a supported, 1.4 mm thick high-flux Pd membrane using feeds obtained by ATR of natural gas in the temperature range most suitable for operation of WGS Pd membrane reactors, that is, above 350°C [178]. CO conversion remained above thermodynamic equilibrium up to feed space velocities of 9100 L/kg/h at 350°C, $P_{total} = 1.2$ MPa, and steam-to-carbon ratio S/C = 3, but H_2 recovery decreased from 84.8% at GHSV = 4050 kg^{-1} h^{-1} to 48.7% at the highest space velocity. This rapid decline of separation performance was explained by slow H_2 diffusion through the catalyst bed.

Li et al. studied the WGS reaction in a bench-scale membrane reactor, where three supported Pd membranes of 44 cm in length and about 6 mm in thickness filled with noble metal shift catalyst were used, reaching a total membrane surface area of 580.6 cm^2 [179]. The WGS reaction was studied with the syngas mixture consisting of 4.0% CO, 19.2% CO_2, 15.4% H_2O, 1.2% CH_4, and 60.1% H_2, at 400°C, p_{feed} = 2-3.5 MPa, and p_{perm} = 1.5 MPa. High reaction pressure and high permeation of Pd membranes allowed for nearly complete CO conversion and H_2 recovery. During 27 days of stable operation both the membranes and the membrane reactor demonstrated a good chemical and mechanical stability under the investigated conditions.

Liguori et al. investigated the performances of WGS reaction carried out in a porous stainless steel supported Pd-based membrane reactor [180]. A 20 μm thick Pd-based membrane prepared by electroless plating and a HT Fe–Cr based WGS catalyst have been used. The influence of reaction pressure (0.7–1.1 MPa) and GHSV (3450–14,000 h^{-1}) were investigated at steam to carbon ratios from 1/1 to 4/1 at a reaction temperature of 390°C. They used a gas mixture consisting of 8% CO, 32% H_2O, 24% CO_2, and 36% H_2. At GHSV = 3450 h^{-1} and 1.1 MPa, almost 80% CO conversion, 70% hydrogen recovery with a hydrogen permeate purity of around 97% was reached. Membrane fouling due to coke coverage on the membrane surface was fully reversible. After 700 h of tests in WGS mixtures, H_2 permeation and both H_2/He and H_2/N_2 ideal selectivities remained unchanged.

While most experimental studies have been performed in reactors with small membrane area, Catalano et al. used in their study a reactor with composite Pd membranes with a surface area of about 0.02 m^2, thus addressing the scaling up of the process [181]. Two thin, δ < 10 μm, defect-free composite membranes were prepared by electroless plating on porous stainless steel tubular supports. They were tested under WGS reaction conditions using a large scale WGS-CMR testing rig consisting of a tube-and-shell reactor, with the annular space surrounding the palladium selective membrane filled with commercial HT iron–chrome oxide shift catalyst. A syngas consisting of 40% H_2, 42.2% CO, and 17.8% CO_2 and steam to carbon ratio varying between 2.5 and 3.5 was fed to the WGS catalytic membrane reactor with a total flow rate up to 1.5 m$_N^3$/h, ≤2 MPa, and 420–440°C. CO conversions higher than the equilibrium conversions were obtained within the entire GHSV range considered in the present study, for pressures between 0.7 and 2 MPa. At a relatively low feed flow rate, GHSV = 1130 h^{-1}, a maximum CO conversion of 98.1% was achieved, with a hydrogen recovery of 81.5% at 440°C. On the other hand, at the highest GHSVs, the system appeared to be limited by the activity of the ferrochrome catalyst. At 2 MPa in the retentate side, 440°C, and GHSV = 5650 h^{-1}, the hydrogen production rate was found to be 5.6 m$_N^3$/day. High hydrogen purity in excess of 99.2% was achieved also in experiments performed with a retentate pressure of 2 MPa.

REFERENCES

[1] Bocci E, Sisinni M, Moneti M, Vecchione L, Di Carlo A, Villarini M. State of art of small scale biomass gasification power systems: a review of different technologies. Energy Procedia 2014;45:247–56.

[2] Li CZ. Importance of volatile-char interactions during the pyrolysis and gasification of low-rank fuels—a review. Fuel 2013;112:609–23.

[3] Bui T, Loof R, Bhattacharya SC. Multi-stage reactor for thermal gasification of wood. Energy 1994;19:397–404.

[4] Bhattacharya SC, Hla SS, Pham H-L. A study on a multi-stage hybrid gasifier-engine system. Biomass Bioenergy 2001;21:445–60.

[5] Henriksen U, Ahrenfeldt J, Jensen TK, Gobel B, Bentzen JD, Hindsgaul C, et al. The design, construction and operation of a 75 kW two-stage gasifier. Energy 2006;31:1542–53.

[6] Cao Y, Wang Y, Riley JT, Pan W-P. A novel biomass air gasification process for producing tar-free higher heating value fuel gas. Fuel Process Technol 2006;87:343–53.

[7] Jaojaruek K, Jarungthammachote S, Gratuito MKB, Wongsuwan H, Homhual S. Experimental study of wood downdraft gasification for an improved producer gas quality through an innovative two-stage air and premixed air/gas supply approach. Bioresour Technol 2011;102:4834–40.

[8] Sulc J, Stojdl J, Richter M, Popelka J, Svoboda K, Smetana J, et al. Biomass waste gasification—can be the two stage process suitable for tar reduction and power generation? Waste Manage 2012;32:692–700.

[9] Raman P, Ram NK, Gupta R. A dual fired downdraft gasifier system to produce cleaner gas for power generation: design, development and performance analysis. Energy 2013;54:302–14.

[10] Galindo AL, Lora ES, Andrade RV, Giraldo SY, Jaen RL, Cobas VM. Biomass gasification in a downdraft gasifier with a two-stage air supply: effect of operating conditions on gas quality. Biomass Bioenergy 2014;61:236–44.

[11] De Capraiis B, De Fillipis P, Scarsella Petrullo A, Palma V. Biomass gasification and tar reforming in a two-stage reactor. Energy Procedia 2014;61:1071–4.

[12] Gomez-Barea A, Leckner B, Villanueva-Perales A, Nilsson S, Fuentes-Cano D. Improving the performance of fluidized bed biomass/waste gasifiers for distributed electricity: a new three-stage gasification system. Appl Therm Eng 2013;50:1453–62.

[13] <http://www.pyroneer.com/en> [accessed 31.07.15].

[14] Hofmann P, Schweiger A, Fryda L, Panopoulos KD, Hohenwarter U, Bentzen JD, et al. High temperature electrolyte supported Ni-GDC/YSZ/LSM SOFC operation on two-stage Viking gasifier product gas. J Power Sources 2007;173:357–66.

[15] Ahrenfeldt J, Thomsen TP, Henriksen U, Clausen LR. Biomass gasification cogeneration—a review of state of the art technology and near future perspectives. Appl Therm Eng 2013;50:1407–17.

[16] Wang Z, He T, Qin J, Wu J, Li J, Zi Z, et al. Gasification of biomass with oxygen-enriched air in a pilot scale two-stage gasifier. Fuel 2015;150:386–93.

[17] Nilsson S, Gomez-Barea A, Fuentes-Cano D, Ollero P. gasification of biomass and waste in a staged fluidized bed gasifier: modeling and comparison with one-stage units. Fuel 2012;97:730–40.

[18] Gomez-Barea A, Ollero P, Leckner B. Optimization of char and tar conversion in fluidized bed biomass gasifiers. Fuel 2013;103:42–52.

[19] Gomez-Barea A, Nilsson S, Vidal-Barrero F, Campoy M. Devolatilization of biomass and waste in fluidized bed. Fuel Process Technol 2010;91:1624–33.

[20] Nilsson S, Gomez-Barea A, Fuentes-Cano D. Gasification reactivity of char from dried sewage sludge in a fluidized bed. Fuel 2012;92:346–53.

[21] Gomez-Barea A, Fuentes-Cano D, Nilsson S, Tirado J, Ollero P. Fluid-dynamics of a cold model of a fluidized bed gasification system with reduced tar content. In proceedings of the 18th European biomass conference 2010, Lyon. p. 669–72.

[22] Stoholm P, Cramer J, Nielsen RG, Sander B, Ahrenfeldt J, Henriksen UB. The low temperature CFB gasifier—$100\,kW_{th}$ tests on straw and new $6\,MW_{th}$ demonstration plant. In proceedings of the 18th European biomass conference 2010, Lyon. p. 619–23.

[23] Zwart R, van der Heijden S, Emmen R, Bentzen JD, Ahrenfeldt J, Stoholm P, Krogh J. Tar removal from low-temperature gasifiers, ECN report 2010. <https://www.ecn.nl/publications//E/2010/ECN-E--10-008.pdf> [accessed 31.07.15].

[24] Thomsen TP, Ahrenfeldt J, Thomsen ST. Assessment of a novel alder biorefinery concept to meet demands of economic feasibility, energy production and long term environmental sustainability. Biomass Bioenergy 2013;53:81–94.

[25] Thomsen TP, Ravenni G, Holm JK, Ahrenfeldt J, Hauggaard-Nielsen H, Henriksen UB. Screening of various low-grade biomass materials for low temperature gasification: method development and application. Biomass Bioenergy 2015;79:128–44.

[26] Wolf B, Meyer B. Verfahrenstechnik und Hauptausrüstung der mehrstufigen Vergasung von Kohle und Biomasse nach dem Carbo-V-Verfahren. DGMK-Tagungsbericht 2000. p. 205–12. ISBN:3-931850-65-X.

[27] Vogels J. Industrial scale hydrogen production from biomass via Choren's unique Carbo-V-process. In: Stolten D, Grube T, editors. 18th world hydrogen energy conference 2010, p. 375–80. <http://juser.fz-juelich.de/record/135494/files/ HP4b_2_Vogels_rev0604.pdf>.

[28] Althapp A. Kraftstoffe aus Biomass emit dem Carbo-V Vergasungsverfahren. <http://www.fvee.de/fileadmin/publikationen/Workshopbaende/ws2003-2/ws2003-2_05_02.pdf> [accessed 11.07.13].

[29] Blades T, Rudloff M, Schulze O. Sustainable SunFuel from Choren's Carbo-V process, ISAF XV, San Diego; 2005. <http://www.eri.ucr.edu/ISAFXVCD/ISAFXVAF/SSFCCVP.pdf> [accessed 31.07.15].

[30] <http://www.erneuerbareenergien.de/choren-meldet-insolvenz-an/150/482/31455/> [accessed 31.07.15].

[31] <http://www.heise.de/tr/artikel/Der-Sprit-ist-aus-1726672.html> [accessed 31.07.15].

[32] Rapier R. What happened at Choren? <http://www.theenergycollective.com/robertrapier/60963/what-happened-choren> [accessed 31.07.15].

[33] <http://www.the-linde-group.com/en/news_and_media/press_releases/news_120209.html> [accessed 31.07.15].

[34] Linde AG. Financial report 2013.

[35] <http://www.greencarcongress.com/2013/01/forest-20130125.html> [accessed 31.07.15].

[36] <http://www.vapo.fi/en/media/news/1997/vapo_oy_freezes_the_kemi_biodiesel_project> [accessed 31.07.15].

[37] Henrich E. The status of the FZK concept of biomass gasification. 2nd European summer school on renewable motor fuels, 29–31 August 2007, Warsaw, Poland.

[38] Dahmen N, Dinjus E, Henrich E. The Karlsruhe process bioliq—synthetic fuels from the biomass Wengenmayr R, Bührke T, editors. Renewable energy. Germany: Wiley-VCH; 2008. p. 61–5.

[39] Trippe F, Fröhling M, Schultmann F, Stahl R, Henrich E. Techno-economic assessment of gasification as a process step within biomass-to-liquid (BtL) fuel and chemicals production. Fuel Process Technol 2011;92:2169–84.

[40] Haro P, Trippe F, Stahl R, Henrich E. Bio-syngas to gasoline and olefins via DME—a comprehensive techno-economic assessment. Appl Energy 2013;108:54–65.

[41] <http://www.bioliq.de> [accessed 01.07.13].

[42] Bridgewater AV. Review of fast pyrolysis of biomass and product upgrading. Biomass Bioenergy 2012;38:68–94.

[43] Jakobs T, Djordjevic N, Fleck S, Mancini M, Weber R, Kolb T. Gasification of high viscous slurry R&D on atomization and numerical simulation. Appl Energy 2012;93:449–56.

[44] Leibold H, Hornung A, Seifert H. HTHP syngas cleaning concept of two stage biomass gasification for FT synthesis. Powder Technol 2008;180:265–70.

[45] Leibold H, Mai R, Linek A, Stöhr J, Seifert H. HTHP syngas cleaning in the bioliq process. 5th international Freiberg conference on IGCC & XtL technologies 2012, Leipzig. <http://www.gasification-freiberg.org> [accessed 01.07.13].

[46] Heidenreich S. Hot gas filtration: a review. Fuel 2013;104:83–94.

[47] Qi Z, Jie C, Tiejun W, Ying X. Review of biomass pyrolysis properties and upgrading research. Energy Convers Manage 2007;48:87–92.

[48] Oasmaa A, Czernik S. Fuel oil quality of biomass pyrolysis oils-state of the art for the end users. Energy Fuels 1999;13:914–21.

[49] Corella J, Toledo JM, Molina-Cuberos GJ. A review on dual fluidized bed biomass gasifiers. Ind Eng Chem Res 2007;46:6831–9.

[50] Kaushal P, Tyagi R. Steam assisted biomass gasification—an overview. Can J Chem Eng 2011;90:1043–58.

[51] Goransson K, Soderlind U, He J, Zhang W. Review of syngas production via biomass DFBGs. Renew Sustain Energy Rev 2011;15:482–92.

[52] Pröll T. Potenziale d. Wirbelschichtdampfvergasung fester Biomasse—Modellierung u. Simulation auf Basis der Betriebserfahrungen am Biomassekraftwerk Güssing, PhD Thesis, TU Wien, 2004.

[53] Hofbauer H, Rauch R, Loeffler G, Kaiser S, Fercher E, Tremmel H. Six years experience with the FICFB-gasification process. In: Proceedings of 12th European conference on biomass for energy 2002, Amsterdam, The Netherlands.

[54] Hofbauer H, Knoef H. Success stories in biomass gasification. In: Handbook biomass gasification, BTG, 2005, 115–61.

[55] Penthor S, Mayer K, Kern S, Kitzler H, Wöss D, Pröll T, et al. Chemical-looping combustion of raw syngas from biomass steam gasification—coupled operation of two dual fluidized bed pilot plants. Fuel 2014;127:178–85.

[56] Kirnbauer F, Hofbauer H. The mechanism of bed material coating in dual fluidized bed biomass steam gasification plants and its impact on plant optimization. Powder Technol 2013;245:94–104.

[57] Schmid JC, Müller S, Hofbauer H. First scientific results with the novel dual fluidized bed gasification test facility at TU Vienna. In: Proceedings of the 24th European biomass conference & exhibition 2016, Amsterdam, The Netherlands, 6–9 June 2016.

[58] Farris M, Paisley MA, Irving J, Overend RP. The biomass gasification process by Battelle/Ferco: design, engineering, construction, and start-up. <http://www.gasification.org/uploads/eventLibrary/gtc9823.pdf> [accessed 30.10.15].

[59] Paisley MA, Overend RP. The SilvaGas process from future energy resources—a commercialization success. In: Proceedings of 12th European conference and technology exhibition on biomass for energy, industry and climate protection, 17–21 June, 2002, Amsterdam, The Netherlands.

[60] <http://www.rentechinc.com/energy-technologies.php> [accessed 03.11.15].

[61] <http://www.biofuelsdigest.com/bdigest/2013/03/01/rentech-to-close-colorado-demo-unit-drop-advanced-biofuels-rd-activities/> [accessed 03.11.15].

[62] Paisley MA. Small scale high throughput biomass gasification system and method. US Patent 6,613,111; 2003.

[63] van der Drift A, van der Meijden CM, Boerrigter H. MILENA gasification technology for high efficient SNG production from biomass. Proceedings of 14th European biomass conference & exhibition, 17–21 October 2005, Paris, France.

[64] van der Meijden CM, van der Drift A, Vreugdenhil BJ. Experimental results from the allothermal biomass gasifier MILENA. In: Proceedings of 15th European biomass conference & exhibition, 7–11 May 2007, Berlin, Germany.

[65] van der Meijden CM, Veringa HJ, van der Drift A, Vreugdenhil BJ. The $800\,kW_{th}$ allothermal biomass gasifier MILENA. In: Proceedings of 16th European biomass conference & exhibition, 2–6 June 2008, Valencia, Spain.

[66] <http://www.royaldahlman.com/renewable/news/press-release-china/> [accessed 30.10.15].

[67] <http://www.royaldahlman.com/renewable/news/commercial-milena-gasifier-under-construction/> [accessed 30.10.15].

[68] <http://www.milenatechnology.com> [accessed 30.10.15].

[69] <http://www.ieatask33.org/app/webroot/files/file/2015/Ponferrada/WS/Drift.pdf> [accessed 30.10.15].

[70] Foscolo PU, Germana A, Jand N, Rapagna S. Design and cold model testing of a biomass gasifier consisting of two interconnected fluidized beds. Powder Technol 2007;173: 179–88.

[71] Di Felice R, Rapagna S, Foscolo PU. Dynamic similarity rules: validity check for bubbling and slugging fluidized beds. Powder Technol 1992;71:281–7.

[72] Nguyen TDB, Seo MW, Lim Y, Song B, Kim S. CFD simulation with experiments in a dual circulating fluidized bed gasifier. Comput Chem Eng 2012;36:48–56.

[73] Bidwe AR, Hawthorne C, Xizhi Y, Dieter H, Scheffknecht G. Cold model study of a dual fluidized bed system for the gasification of solid fuel. Fuel 2014;127:151–60.

[74] Shrestha S, Ali BS, Jan BM, Hamid MDB, El Sheikh K. Hydrodynamic characteristics in cold model of dual fluidized bed gasifiers. Powder Technol 2015;286:246–56.

[75] Lim MT, Saw W, Pang S. Effect of fluidizing velocity on gas bypass and solid fraction in a dual fluidized bed gasifier and a cold model. Particuology 2015;18:58–65.

[76] Pasteiner H, Schmid JC, Müller S, Hofbauer H. Cold flow investigations on a novel dual fluidized bed steam gasification test plant. In: Presented at the minisymposium chemical engineering 2015, 14–15June. Austria: University of Natural Science Vienna (BOKU).

[77] Miccio F, Ruoppolo G, Kalisz S, Andersen L, Morgan TJ, Baxter D. Combined gasification of coal and biomass in internal circulating fluidized bed. Fuel Process Technol 2012;95:45–54.

[78] Simanjuntak JP, Zainal ZA. Experimental study and characterization of a two-compartment cylindrical internally circulating fluidized bed gasifier. Biomass Bioenergy 2015;77:147–54.

[79] Di Felice L, Courson C, Jand N, Gallucci K, Foscolo PU, Kiennemann A. Catalytic biomass gasification: simultaneous hydrocarbons steam reforming and CO2 capture in a fluidised bed reactor. Chem Eng J 2009;154:375–83.

[80] Marquard-Möllenstedt T, Zuberbuehler U, Specht M. Transfer of the absorption enhanced reforming (AER)-process from the pilot scale to an 8 MWTH gasification plant in Guessing, Austria. In: Proceedings of 16th European biomass conference and exhibition from research to industry and markets, 2–6 June 2008, Valencia, Spain. p. 684–9.

[81] Pfeifer C, Puchner B, Hofbauer H. Comparison of dual fluidized bed steam gasification of biomass with and without selective transport of CO2. Chem Eng Sci 2009;64:5073–83.

[82] Koppatz S, Pfeifer C, Rauch R, Hofbauer H, Marquard-Moellenstedt T, Specht M. H_2 rich product gas by steam gasification of biomass with in situ CO_2 absorption in a dual fluidized bed system of 8 MW fuel input. Fuel Process Technol 2009;90:914–21.

[83] Hejazi B, Grace JR, Bi X, Mahecha-Botero A. Steam gasification of biomass coupled with lime-based CO2 capture in a dual fluidized bed reactor: a modeling study. Fuel 2014;117:1256–66.

[84] Devi L, Ptasinski KJ, Janssen FJJG. A review of the primary measures for tar elimination in biomass gasification processes. Biomass Bioenergy 2003;24:125–40.

[85] Anis S, Zainala ZA. Tar reduction in biomass producer gas via mechanical, catalytic and thermal methods: a review. Renew Sustain Energy Rev 2011;15:2355–77.

[86] Han J, Kim H. The reduction and control technology of tar during biomass gasification/pyrolysis: an overview. Renew Sustain Energy Rev 2008;12:397–416.

[87] Houben MP, de Lange HC, van Steenhoven AA. Tar reduction through partial combustion of fuel gas. Fuel 2005;84:817–24.

[88] Ilham K, Brandin J, Sanati M. Shift catalysis in biomass generated synthesis gas. Top Catal 2007;45:31–7.

[89] Zhang Y, Kajitani S, Ashizawa M, Oki Y. Tar destruction and coke formation during rapid pyrolysis and gasification of biomass in a drop-tube furnace. Fuel 2010;89:302–9.

[90] van der Hoeven TA, de Lange HC, van Steenhoven AA. Analysis of hydrogen-influence on tar removal by partial oxidation. Fuel 2006;85:1101–10.

[91] Svensson H, Tuna P, Hulteberg C, Brandin J. Modeling of soot formation during partial oxidation of producer gas. Fuel 2013;106:271–8.

[92] Ahrenfeldt J, Egsgaard H, Stelte W, Thomsen T, Henriksen UB. The influence of partial oxidation mechanisms on tar destruction in two stage biomass gasification. Fuel 2013;112:662–80.

[93] Brandt P, Larsen E, Henriksen U. High tar reduction in a two-stage gasifier. Energy Fuels 2000;14:816–9.

[94] Su Y, Luo Y, Wu W, Zhang Y. Experimental and numerical investigation of tar destruction oxidation environment. Fuel Process Technol 2011;92:1513–24.

[95] Weston PM, Sharifi V, Swithenbank J. Destruction of Tar in a Novel Coandă Tar Cracking System. Energy Fuels 2014;28:1059–65.

[96] Burhenne L, Rochlitz L, Lintner C, Aicher T. Technical demonstration of the novel Fraunhofer ISE biomass gasification process for the production of a tar-free synthesis gas. Fuel Process Technol 2013;106:751–60.

[97] Carnö J, Berg M, Järås S. Catalytic abatement of emissions from small-scale combustion of wood: a comparison of the catalytic effect in model and real flue gases. Fuel 1996;75:959–65.

[98] Miyazawa T, Kimura T, Nishikawa J, Kunimori K, Tomishige K. Catalytic properties of Rh/CeO_2/SiO_2 for synthesis gas production from biomass by catalytic partial oxidation of tar. Sci Technol Adv Mater 2005;6:604–14.

[99] Tomishige K, Miyazawa T, Kimure T, Kunimori K. Novel catalyst with high resistance to sulfur for hot gas cleaning at low temperature by partial oxidation of tar derived from biomass. Catal Commun 2005;6:37–40.

[100] Miyazawa T, Kimura T, Nishikawa J, Kado S, Kunimori K, Tomishige K. Catalytic performance of supported Ni catalysts in partial oxidation and steam reforming of tar derived from the pyrolysis of wood biomass. Catal Today 2006;115:254–62.

[101] Wang C, Wang T, Ma L, Gao Y, Wu C. Partial oxidation reforming of biomass fuel gas over nickel-based monolithic catalyst with naphthalene as model compound. Korean J Chem Eng 2008;25:738–43.

[102] Herrmann S. Selektivoxidation von Naphthalin in CO/H2-Mischungen an Mo/V/W-Mischoxiden—Ein Beitrag zur Biomassevergasung. PhD-thesis, TU Darmstadt, Germany; 2007.

[103] Ammendola P, Lisi L, Piriou B, Ruoppolo G. Rh-perovskite catalysts for conversion of tar from biomass pyrolysis. Chem Eng J 2009;154:361–8.

[104] Böhning D. Katalytisch partielle Oxidation polyzyklischer aromatischer Kohlenwasserstoffe in Brenngasen aus der Biomassevergasung—Modellierung und experimentelle Untersuchungen. PhD-thesis, TU Dresden, Germany; 2010.

[105] Ma M, Müller M, Richter J, Kriegel R, Böhning D, Beckmann M, et al. Investigation of combined catalyst and oxygen carrier systems for the partial oxidation of naphthalene as model tar from biomass gasification. Biomass Bioenergy 2013;53:65–71.

[106] Hossain MM, de Lasa HI. Chemical-looping combustion (CLC) for inherent CO_2 separations—a review. Chem Eng Sci 2008;63:4433–51.

[107] Ma M, Müller M. Investigation of various catalysts for partial oxidation of tar from biomass gasification. Appl Catal A 2015;493:121–8.

[108] Böhning D, Beckmann M, Kriegel R, Richter J, Ma M, Müller M. Kombiniertes Katalysator- und Sauerstoffträgersystem zur Aufbereitung teerhaltiger Brenngase aus der Biomassevergasung via partieller Oxidation. Chem Ing Tech 2015;87:457–65.

[109] Xu SJ, Thomson WJ. Oxygen permeation rates through ion-conducting perovskite membranes. Chem Eng Sci 1999;54:3839–50.

[110] Teraoka Y, Zhang HM, Furukawa S, Yamazoe N. Oxygen permeation through perovskite-type oxides. Chem Lett 1985;14:1743–6.

[111] Teraoka Y, Zhang HM, Okamoto K, Yamazoe N. Mixed ionic-electronic conductivity of $La_{1-x}Sr_xCo_{1-y}Fe_yO_{3-\delta}$ perovskite-type oxides. Mater Res Bull 1988;23:51–8.

[112] VITO (Vision on technology). Catalogue Materials Technology. <www.vito.be> [accessed 30.09.15].

[113] Wei Y, Yang W, Caro J, Wang H. Dense ceramic oxygen permeable membranes and catalytic membrane reactors. Chem Eng J 2013;220:185–203.

[114] Kalisz S, Pronobis M, Baxter D. Co-firing of biomass waste-derived syngas in coal power boilers. Energy 2008;33:1770–8.

[115] Basu P, Butler J, Leon MA. Biomass co-firing options on the emission reduction and electricity generation costs in coal-fired power plants. Renew Energy 2011;36:282–8.

[116] Al-Mansour F, Zuwala J. An evaluation of biomass co-firing in Europe. Biomass Bioenergy 2010;34:620–9.

[117] Saidur R, Abdelaziz EA, Demirbas A, Hossain MS, Mekhilef S. A review on biomass as a fuel for boilers. Renew Sustain Energy Rev 2011;15:2262–89.

[118] Willeboer W. Biomass co-firing in present and future power blocks. IEA Bioenergy Task 32 Workshop October 2008, Geertruidenberg. <http://www.ieabcc.nl/workshops/task32_Amsterdam2008/cofiring/Wim%20Willeboer.pdf> [accessed 10.11.15].

[119] Kristensen T. Fuel flexibility in the new coal, biomass and oil fired CHP plant Amager Unit 1 in Copenhagen. IEA Bioenergy Task 32 Workshop June 2009, Hamburg. <http://www.ieabcc.nl/workshops/task32_Hamburg2009/cofiring/02%20kristensen.pdf> [accessed 10.11.15].

[120] Kiel J. Biomass co-firing in high percentage—opportunities in conventional and advanced coal-fired power plants. IEA Bioenergy Task 32 Workshop October 2008, Geertruidenberg. <http://www.ieabcc.nl/workshops/task32_Amsterdam2008/cofiring/Jaap%20Kiel.pdf> [accessed 10.11.15].

[121]   Nielsen HP, Baxter LL, Sclippab G, Morey C, Frandsen FJ, Dam-Johansen K. Deposition of potassium salts on heat transfer surfaces in straw-fired boilers: a pilot-scale study. Fuel 2000;79:131–9.

[122]   Nielsen HP, Frandsen FJ, Dam-Johansen K, Baxter LL. The implications of chlorine-associated corrosion on the operation of biomass-fired boilers. Prog. Energy Combust Sci. 2000;26:283–98.

[123]   Dong C, Yang Y, Yang R, Zhang J. Numerical modeling of the gasification based biomass co-firing in a 600 MW pulverized coal boiler. Appl Energy 2010;87:2834–8.

[124]   Nieminen J, Kivela M. Biomass CFB gasifier connected to a 350 MW$_{th}$ steam boiler fired with coal and natural gas—Thermie demonstration project in Lahti in Finland. Biomass Bioenergy 1998;15:251–7.

[125]   Raskin N, Palonen J, Nieminen J. Power boiler fuel augmentation with a biomass fired atmospheric circulating fluid-bed gasifier. Biomass Bioenergy 2001;20:471–81.

[126]   Anttikoski T, Palonen J, Eriksson T. Foster Wheeler biomass gasifier experiences from Lahti & Ruien and further cases for difficult biomass & RDF gasification. IEA bioenergy workshop: co-utilisation of biomass with fossil fuels, 25 May 2005. Copenhagen, Denmark.

[127]   Palonen J, Anttikoski T, Eriksson T. The Foster Wheeler gasification technology for biofuels: refuse-derived fuel (RDF) power generation, Power-Gen Europe, May 30–June 1, 2006, Kölnmesse, Cologne, Germany.

[128]   Jouret N, Helsen L, Van den Bulck E. Study of the wood gasifier at the power plant of Electrabel-Ruien. In: Proceedings of the European combustion meeting, 3–6 April 2005, Louvain-La-Nouve, Belgium.

[129]   Ryckmans Y, Van den Spiegel F. biomass gasification and use of the syngas as an alternative fuel in a Belgian coal-fired boiler. <http://www.laborelec.com/docs/articles/lbe_art_COMB006_uk.pdf> [accessed 10.11.15].

[130]   Willeboer W. The Amer demolition wood gasification project. Biomass Bioenergy 1998;15:245–9.

[131]   Vollebregt E. Essent's experiences with large scale biomass co-firing. <http://www.essent.eu/content/Images/90686_06_Large%20scale%20biomass%20firing%20Essent_Etienne%20Vollebregt.pdf> [accessed 10.11.15].

[132]   Mory A, Zotter T. EU-demonstration project BioCoComb for biomass gasification and co-combustion of the product-gas in a coal-fired power plant in Austria. Biomass Bioenergy 1998;15:239–44.

[133]   Granatstein DL. Case study on BioCoComb biomass gasification project Zeltweg power station, Austria, IEA Bioenergy—Task 36 Report 2002. <http://www.ieabioenergytask36.org/Publications/2001-2003/Case_Studies/Case_Study_on_BioCoComb_Biomass_Gasification_Project.pdf> [accessed 10.11.15].

[134]   <http://www.pyroneer.com/en/demonstration-plant/asnaes-power-plant> [accessed 10.11.15].

[135]   <http://www.perinijournal.com/Items/en-US/Articoli/Notes/Worlds-largest-biomass-gasification-plant-inaugurated-in-Vaasa-plant-supplied-by-Metso> [accessed 10.11.15].

[136]   Isaksson J. Commercial scale gasification to replace fossil flue in power generation—Vaskiluodon Voima 140 MW CFB gasification project. In: IEA bioenergy conference, 27–29 October 2015, Berlin, Germany. <https://ieabioenergy2015.org/fileadmin/veranstaltungen/2015/IEA_Bioenergy_Conference/S01-2_Isaksson.pdf> [accessed 10.11.15].

[137]   Franz S. Energetische und ökonomische Analyse der Kohlenstoffdioxidrückhaltung in Kohlevergasungskraftwerken mittels Polymer- und Keramikmembranen. PhD-thesis, Ruhr-Universität Bochum; 2012.

[138] D'Alessandro DM, Smit B, Long JR. Carbon dioxide capture: prospects for new materials. Angew Chem Int Ed 2010;49:6058–82.

[139] Shao L, Low BT, Chung TS, Greenberg AR. Polymeric membranes for the hydrogen economy: contemporary approaches and prospects for the future. J Membrane Sci 2009;327:18–31.

[140] Zou J, Huang J, Ho WSW. CO_2-selective water gas shift membrane reactor for fuel cell hydrogen processing. Ind Eng Chem Res 2006;46:2272–9.

[141] Zou J, Ho WSW. CO_2-selective polymeric membranes containing amines in crosslinked poly(vinyl alcohol). J Membrane Sci 2006;286:310–21.

[142] Shimekit B, Mukhtar H, Ahmad F, Maitra S. Ceramic membranes for the separation of carbon dioxide—a review. Trans Indian Ceram S 2009;68:115–38.

[143] Scholes CA, Smith KH, Kentish SE, Stevens GW. CO_2 capture from pre-combustion processes—strategies for membrane gas separation. Int J Greenh Gas Con 2010;4:739–55.

[144] Wade J, Lackner K. High temperature CO_2 membrane separation: enabling new carbon capture and coal conversion strategies. In: Proceedings of the 39th international technical conference on clean coal & fuel systems 2014, Clearwater (FL).

[145] Gallucci F, Fernandez E, Corengia P, van Sint Annaland M. Recent advances on membranes and membrane reactors for hydrogen production. Chem Eng Sci 2013;92:40–66.

[146] Kluiters SCA. Status review on membrane systems for hydrogen separation. Intermediate Report EU Project MIGREYD NNE5-2001-670, December 2004.

[147] Liu K, Song C, Subramani V. Hydrogen and syngas production and purification technologies. New Jersey: John Wiley & Sons Incorporation; 2010.

[148] Chung TS, Shao L, Tin PS. Surface modification of polyimide membranes by diamines for H_2 and CO_2 separation. Macromol Rapid Comm 2006;27:998–1003.

[149] Pesiri DR, Jorgensen B, Dye RC. Thermal optimization of polybenzimidazole meniscus membranes for the separation of hydrogen, methane, and carbon dioxide. J Membrane Sci 2003;218:11–18.

[150] Phair JW, Donelson R. Developments and design of novel (non-palladium-based) metal membranes for hydrogen separation. Ind Eng Chem Res 2006;45:5657–74.

[151] Uemiya S, Matsuda T, Kikuchi E. Hydrogen permeable palladium-silver alloy membrane supported on porous ceramics. J. Membrane Sci 1991;56:315–25.

[152] Mardilovich PP, She Y, Ma YH, Rei MH. Defect-free palladium membranes on porous stainless-steel support. AIChE J 1998;44:310–22.

[153] Dittmeyer R, Höllein V, Daub K. Membrane reactors for hydrogenation and dehydrogenation processes based on supported palladium. J Mol Catal A-Chem 2001;173:135–84.

[154] Lin YM, Rei MH. Separation of hydrogen from the gas mixture out of catalytic reformer by using supported palladium membrane. Sep Purif Technol 2001;25:87–95.

[155] Gao H, Lin YS, Li Y, Zhang B. Chemical stability and its improvement of palladium-based metallic membranes. Ind Eng Chem Res 2004;43:6920–30.

[156] Wang D, Tong J, Xu H, Matsumura Y. Preparation of palladium membrane over porous stainless steel tube modified with zirconium oxide. Catal Today 2004;93:689–93.

[157] Robertson WM. Hydrogen permeation, diffusion and solution in nickel. Z Metallkd 1973;64:436–43.

[158] Steward SA. Review of hydrogen isotope permeability through materials. Livermore (CA): Lawrence Livermore National Laboratory, University of California; 1983.

[159] Ozaki T, Zhang Y, Komaki M, Nishimura C. Hydrogen permeation characteristics of V–Ni–Al alloys. Int J Hydrogen Energy 2003;28:1229–35.

[160] Daw M, Baskes M. Semiempirical, quantum mechanical calculation of hydrogen embrittlement in metals. Phys Rev Lett 1983;50:1285–8.

[161] Musket RG. Effects of contamination on the interaction of hydrogen gas with palladium: a review. J Less Common Metals 1976;45:173–83.

[162] Asaeda M, Yamasaki S. Separation of inorganic/organic gas mixtures by porous silica membranes. Sep Purif Technol 2001;25:151–9.

[163] Aoki K, Kusakabe K, Morooka S. Gas permeation properties of A-type zeolite membrane formed on porous substrate by hydrothermal synthesis. J Membrane Sci 1998;141:197–205.

[164] Petersen J, Matsuda M, Haraya K. Capillary carbon molecular sieve membranes derived from Kapton for high temperature gas separation. J Membrane Sci 1997;131:85–94.

[165] Higuchi T, Tsukamoto T, Sata N, Hattori T, Yamaguchi S, Shin S. Electronic structure of protonic conductor $SrCeO_3$ by soft-X-ray spectroscopy. Solid State Ionics 2004;175:549–52.

[166] Song SJ, Wachsman ED, Dorris SE, Balachandran U. Defect chemistry modeling of high-temperature proton-conducting cerates. Solid State Ionics 2002;149:1–10.

[167] Guan J, Dorris SE, Balachandran U, Liu M. Transport properties of $BaCe_{0.95}Y_{0.05}O_{3-\delta}$ mixed conductors for hydrogen separation. Solid State Ionics 1997;100:45–52.

[168] Ma GL, Shimura T, Iwahara H. Ionic conduction and nonstoichiometry in $Ba_xCe_{0.90}Y_{0.10}O_{3-\delta}$. Solid State Ionics 1998;110:103–10.

[169] Tetsuo S, Kazumasa E, Hiroshige M, Hiroyasu I. Protonic conduction in Rh-doped $AZrO_3$ (A=Ba, Sr and Ca). Solid State Ionics 2002;149:237–46.

[170] Phair JW, Badwal SPS. Review of proton conductors for hydrogen separation. Ionics 2006;12:103–15.

[171] Haugsrud R. Defects and transport properties in Ln_6WO_{12} (Ln=La, Nd, Gd, Er). Solid State Ionics 2007;178:555–60.

[172] Haugsrud R, Norby T. High-temperature proton conductivity in acceptor-doped $LaNbO_4$. Solid State Ionics 2006;177:1129–35.

[173] Giessler S, Jordan L, Diniz da Costa JC, Lu GQ. Performance of hydrophobic and hydrophilic silica membrane reactors for the water gas shift reaction. Sep Purif Technol 2003;32:255–64.

[174] Van Holt D, Forster E, Ivanova ME, Meulenberg WA, Müller M, Baumann S, et al. Ceramic materials for H_2 transport membranes applicable for gas separation under coal-gasification-related conditions. J Eur Ceram Soc 2014;34:2381–9.

[175] Van Holt D. Keramische Membranen für die H_2-Abtrennung in CO-Shift-Reaktoren. PhD-thesis. Bochum, Germany: Bochum University; 2014.

[176] Forster EMH. Thermochemische Beständigkeit von keramischen Membranen und Katalysatoren für die H_2-Abtrennung in CO-Shift-Reaktoren. PhD-thesis. Aachen, Germany: RWTH Aachen University; 2015.

[177] Augustine AS, Ma YH, Kazantzis NK. High pressure palladium membrane reactor for the high temperature water–gas shift reaction. Int J Hydrogen Energy 2011;36:5350–60.

[178] Bi Y, Xu H, Li W, Goldbach A. Water-gas shift reaction in a Pd membrane reactor over Pt/$Ce_{0.6}Zr_{0.4}O_2$ catalyst. Int J Hydrogen Energy 2009;34:2965–71.

[179] Li H, Pieterse JAZ, Dijkstra JW, Boon J, van den Brink RW, Jansen D. Bench-scale WGS membrane reactor for CO_2 capture with co-production of H_2. Int J Hydrogen Energy 2012;37:4139–43.

[180] Liguori S, Pinacci P, Seelamd PK, Keiskid R, Drago F, Calabrò V, et al. Performance of a Pd/PSS membrane reactor to produce high purity hydrogen via WGS reaction. Catal Today 2012;193:87–94.

[181] Catalano J, Guazzone F, Mardilovich IP, Kazantzis NK, Ma YH. Hydrogen production in a large scale water-gas shift Pd-based catalytic membrane reactor. Ind Eng Chem Res 2013;52:1042–55.

Chapter 6

New and Improved Gasification Concepts

6.1 OXYGEN–STEAM GASIFICATION

The main gasification reactions are endothermic reactions (see chapter: Fundamental Concepts in Biomass Gasification). Therefore, gasification processes require substantial thermal input to operate the reactor. The gasifying agents can be air, oxygen, steam, or their combination. In the case of biomass gasification, usually air is used as an oxidant because of its low-cost and availability. However, product gas from air-blown gasification contains about 40–60 vol.% N_2 and has therefore a low heating value of 3–6.5 MJ/m_N^3 (see Table 2.1). By using oxygen/steam as the gasifying agent the syngas calorific value can be significantly increased to values of 12–17 MJ/m_N^3. Furthermore, a syngas diluted by nitrogen seems to be useful only for electricity production or heat generation but is not suitable for many chemical synthesis processes. Further costly gas separation steps operating high producer gas volumes would be required in the case of direct combination of air-blown gasification and synthesis. Steam is frequently used in combination with air or oxygen since it increases H_2 concentration in the producer gas via the WGS and reforming reactions. Furthermore, it can help to control the temperature in the gasifier due to endothermic steam reforming. Therefore, research on oxygen and steam enriched air-blown gasification, for example, Refs. [1–4], and oxygen/steam-blown gasification of biomass has been performed in recent years. Two examples of the latter will be discussed in detail below, further examples can be found in the literature, for example, Refs. [5–8].

A pilot plant for fluidized bed gasification of biomass with steam–O_2 mixtures was operated at the University of Saragossa, Spain, between 1992 and 1996 [9]. The gasifier had an inner diameter of 15 cm and a height of 3.2 m. It was fed with pine wood chips at flow rates of 5–20 kg/h. The main operating variables studied were gasifier bed temperature (780–890°C), steam to oxygen ratio (2–3 mol/mol), and gasifying agent ($H_2O + O_2$) to biomass ratio (0.6–1.6 kg/kg). For a H_2O/O_2 ratio of 3 the gasifier worked as an autothermal reactor. For gasification with steam and oxygen, the producer gas composition at the gasifier exit was on a dry basis 13–29 vol.% H_2, 30–50 vol.% CO, 14–37 vol.%

Advanced Biomass Gasification. DOI: http://dx.doi.org/10.1016/B978-0-12-804296-0.00006-3

CO_2, 5–7.5 vol.% CH_4, and 2.3–3.8 vol.% C_2-compounds. Thus, a lower heating value (LHV) of 11.4–15.7 MJ/m_N^3 (dry basis) was achieved. The steam content in the raw producer gas was 32–60 vol.%. The gas yield was 0.86–1.2 m_N^3 dry gas/ kg biomass, the char yield 5–20 wt%, and the apparent thermal efficiency (defined as the ratio between LHV of dry product gas and LHV of feed) approximately 60–97%. The main components of the tar produced were phenol, cresol, naphthalene, indene, and toluene. The tar content in the raw gas varied between 2 and 50 g/m_N^3 in the dry gas. Based on the obtained results and experience during the test campaigns the authors considered that the best and recommended operating conditions for the studied process are the following: a gasifier bed temperature between 800°C and 860°C, a H_2O/O_2-ratio of about 3.0 (mol/mol), a ($H_2O + O_2$)/biomass- ratio of 0.8–1.2 (kg/kg), and a gas residence time (in the gasifier bed) of about 2 s. With these conditions a relatively clean gas with a tar content of about 5 g/m_N^3 was obtained. To get a hot fuel gas with a lesser tar content or a higher H_2 content, the group also studied hot gas cleaning and upgrading with calcined dolomite [10] and commercial steam reforming catalysts [11] located downstream of the fluidized bed. Using calcined dolomite H_2 and CO content in the fuel gas increases by 16–23 vol.% and decreases by 15–22 vol.% (dry basis), respectively. Although CH_4 conversion higher than 30 vol.% was not reached, conversion of heavier tars of 90–95 vol.% was obtained with space times (defined as the ratio between the mass of calcined dolomite and the volume flow of the fuel gas) of 0.06–0.15 kgh/m^3. Also the gas yield was increased by 0.15–0.40 m_N^3/kg biomass. Using different commercial nickel-based steam reforming catalysts in combination with a guard bed with a calcined dolomite before the catalytic bed to decrease the tar content in the raw gas below the limit of 2 g of tar/m_N^3, thus avoiding the catalyst deactivation by coke formation, H_2 and CO contents increased by 4–14 and 1–8 vol.% (dry basis), respectively. CO_2 and CH_4 decreased by 0–14, 87–99 vol.% (dry basis), respectively. The steam content decreased by 2-6% at progressively increased temperature. The LHV decreased by 0.3–1.7 MJ/m_N^3, the gas yield increased by 0.1–0.4 m_N^3/kg of biomass, and the apparent thermal efficiency increased by 1–20%.

In recent years, a 100 kW_{th} atmospheric steam-oxygen blown circulating fluidized-bed (CFB) gasifier has been operated at Delft University of Technology, the Netherlands [12–16]. In the experiments several biomass types of both woody and agricultural origin have been used, that is, several types of clean wood, demolition wood, the energy crop species *Miscanthus*, straw, and the agriculture residue Dry Distiller's Grains with Solubles (DDGS). The effects of operational conditions like steam/biomass ratio (SBR), oxygen/biomass stoichiometric ratio (ER), and gasification temperature on the composition distribution of the producer gas and tar formation from these fuels were investigated. Moreover, different bed materials have been applied, namely quartz sand, treated and untreated olivine, and magnesite. During the experiments extensive measurements of gas composition were performed throughout the integrated test rig. The gas characterization included major gas components as well as certain minor species and tar. For a SBR of 4.4–6.8 (mol/mol) and an ER of 0.24–0.38, the producer gas composition at 800–850°C was on a dry basis 19.4–35.9 vol.% H_2, 12.7–33.7 vol.% CO, 30.7–45.3 vol.% CO_2, 5.5–9.5 vol.% CH_4, and 1.4–4.3 vol.% C_2-compounds. The steam content in the raw producer gas was 49–67 vol.%. A fairly high amount of H_2S (~2300 ppmv), COS (~200 ppmv) and trace amounts of methyl mercaptan (<3 ppmv) were obtained from DDGS, which contains the highest amount of sulfur (0.76 wt%) among the investigated biomass types. Due to a relatively high content of K and Cl in herbaceous biomass and DDGS, an alkali-getter (eg, kaolin) was added to avoid bed agglomeration during gasification. The tar content in the raw gas varied between 2 and 50 g/m_N^3 in the dry gas. Using magnesite as bed material, the highest hydrogen yield was achieved among the several tested bed materials. Also the H_2/CO ratio increased from values near or below 1 to 2.3–2.6, which is near the value needed for, for example, Fischer–Tropsch synthesis. Furthermore, the tar content of the raw gas was reduced to values near 2 g/m_N^3. Moreover, magnesite had a positive impact on agglomeration prevention for the agricultural fuels containing alkali and chlorine in the ash. Finally, kaolin proved to be an effective additive to counteract the agglomeration when fuels with high alkali content in the ash are gasified using a bed material which is rich in silica, as is the case with quartz sand and olivine.

The major drawback of oxygen/steam-blown biomass gasification is the need for an air separation unit. While cryogenic air separation is state-of-the-art in large-scale coal gasification, in small- to medium-scale biomass gasification operating at ambient pressure, air separation is mainly realized by selective nitrogen–oxygen sorption systems, such as pressure swing adsorption (PSA), which require air compression. Thus, a gas stream five times greater than the oxygen stream effectively utilized in the gasification process is compressed, even more when considering that the PSA separation efficiency is less than one. This causes a substantial penalty of the whole energy efficiency: about 20% of power generated by the biomass conversion plant would be needed to provide the required oxygen stream.

As alternative to conventional air separation processes, mixed ionic (oxygen ions) and electronic conducting (MIEC) membranes are under investigation for oxygen supply [17].

The use of oxygen transport membranes (OTMs) was already mentioned in Section 5.2.2 in relation to catalytic, chemical looping partial oxidation of tar, as an alternative to oxygen carrier materials (OCMs). A brief account of the OTM oxygen transport mechanism was already provided there. This type of membrane operates usually at temperatures >700°C. The driving force for oxygen permeation is the difference in oxygen partial pressure across the membrane Section 5.2.2[18]. Since the oxygen partial pressure in gasification systems is very low, the air feeding stream does not need to be pressurized. Using MIEC membranes, very pure oxygen can be supplied and the energy demand for oxygen separation can be substantially reduced compared to conventional air separation systems with evident and noticeable advantages especially for small- to medium-scale systems.

There are many research activities on the integration of MIEC membranes in oxyfuel combustion processes, for example, Refs. [19–21], and in coal-fired integrated gasification combined cycle power plants, for example, Ref. [22]. Recently, Gupta et al. tested an OTM at 850–900°C in an experimental coal gasifier [23]. The oxygen flux was stable over 80h and the OTM remained chemically and structurally stable against coal ash. However, integration of MIEC membranes in biomass gasification plants has not been extensively studied or even demonstrated so far.

Recently, Puig-Arnavat et al. modeled different configurations for oxygen production using MIEC membranes where the oxygen partial pressure difference is achieved by creating a vacuum on the permeate side, compressing the air on the feed side, or a combination of the two [24]. The two configurations demonstrating the highest efficiency are then thermally integrated into an oxygen–steam biomass gasification plant. One configuration is exemplarily shown in Fig. 6.1. The energy demand for oxygen production and the membrane area required for a $6\,MW_{th}$ biomass plant were calculated for different operating conditions. Increasing the air feed pressure increases the energy consumption but decreases the membrane area. As an example, for the highest efficiency configuration working at a membrane temperature of 850°C, 6 bar of air feed pressure, and 0.3 bar of permeate side pressure, $150\,m^2$ are needed to generate the oxygen for the plant at an energy consumption of 100 kWh per t_{O2}.

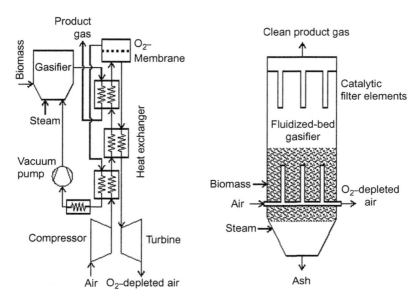

FIGURE 6.1   Concepts for integration of oxygen transport membranes into a gasification process (left) [24] and directly into the gasifier (right) [25].

While Puig-Arnavat et al. [24] investigated only combination and heat integration of the membrane and the gasification process, Antonini et al. proposed the direct integration of a MIEC membrane into a fluidized-bed gasifier [25,26], as illustrated in Fig. 6.1 and already mentioned in Section 4.7. They studied the application of MIEC membrane technology for oxygen separation from air to biomass conversion systems by coupling an oxygen transfer model to a gasification model that considers thermodynamic and kinetic constraints. Numerical evaluations were performed of char partial combustion with oxygen permeated through the membrane, in the gasifier region close to the tubular membrane surface, as a means to provide the necessary input of heat to biomass gasification. To validate the simulation results, experimental oxygen permeation data were taken into account. A bench-scale gasifier (inner diameter: 100 mm) was considered, with a cylindrical pipe (length: 280 mm) immersed vertically inside the fluidized bed, which is made of a layer of perovskite membrane (thickness: 0.3 mm) supported on a ceramic porous structure. The numerical results show that the membrane surface needed to provide the required oxygen flow to the gasifier is small enough to be arranged inside the fluidized bed volume, assuring feasibility of an autothermal process. An estimate of the effect of temperature perturbation due to char combustion at the membrane surface indicates that the associated increase in permeability is substantial. The model is also helpful to optimize the location of the membrane module and evaluate different options. However, more experimental investigations are needed to check the resistance and durability of membrane materials in the gasifier environment.

6.2 CHEMICAL LOOPING GASIFICATION

Gasification with oxygen or oxygen/steam mixtures as the gasifying agent result in a producer gas with higher heating value than producer gas from air-blown gasification as discussed in the chapter "Advanced Process Combination Concepts". However, the supply of pure oxygen by conventional processes, such as cryogenic air separation or PSA, increases the operation costs and decreases significantly the overall efficiency of oxygen gasification in comparison to air gasification.

Chemical looping gasification (CLG) of biomass is a relatively new technology which can provide a solution to the issues mentioned previously. In CLG, lattice oxygen from OCMs is used to provide pure oxygen for the gasification reactions [27]. According to a techno-economic comparison done by Aghabararnejad et al. [28], the capital investment of a CLG unit would be higher than that of a conventional gasification system with a supply of pure oxygen by PSA. However, the annual operating cost of the former would be less so that the difference in investment would be compensated in less than 6 years. This confirms the feasibility of CLG as an alternative to other oxygen-blown gasification processes.

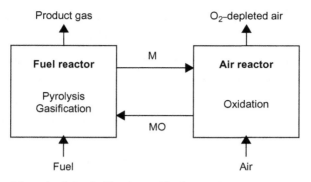

FIGURE 6.2   Schematic of chemical looping gasification.

The concept of CLG is derived from chemical looping combustion (CLC) [29]. The main difference is that the target product of CLG is undiluted syngas instead of heat. Similar to CLC, the CLG process consists of two separate reactors, an air reactor (AR) and a fuel reactor (FR) [30], as illustrated in Fig. 6.2. In the FR, biomass is pyrolyzed at high temperature into gas, tar, and char first and then the pyrolysis products react with the oxygen carriers (OCs). The OCs provide the necessary oxygen and tar and char are partially oxidized into syngas. As a result, the metal oxide, which is most often used as OCM, is reduced to a lower oxidation state. Simultaneously, the OCs can act as catalysts enhancing the decomposition and reformation of tar [31,32]. The following reactions with the OCM occur, where MO denotes the OCM in the higher oxidation state (fresh or after regeneration) and M denotes the OCM in the reduced oxidation state:

$$CO + MO \leftrightarrow CO_2 + M \qquad \Delta H < 0 \text{ kJ/mol} \qquad (6.1)$$

$$H_2 + MO \leftrightarrow H_2O + M \qquad \Delta H < 0 \text{ kJ/mol} \qquad (6.2)$$

$$CH_4 + MO \leftrightarrow 2H_2 + CO + M \qquad \Delta H > 0 \text{ kJ/mol} \qquad (6.3)$$

$$CH_4 + 3MO \leftrightarrow 2H_2O + CO_2 + 3M \qquad \Delta H < 0 \text{ kJ/mol} \qquad (6.4)$$

$$C + MO \leftrightarrow CO + M \qquad \Delta H < 0 \text{ kJ/mol} \qquad (6.5)$$

$$C + 2MO \leftrightarrow CO_2 + M \qquad \Delta H < 0 \text{ kJ/mol} \qquad (6.6)$$

After the reduction process, the reduced OCs are transported into the AR where they are regenerated with air to recover the used lattice oxygen. Since the oxidation reaction is exothermic, the heat required for gasification reactions can be provided by the circulating OC from the AR.

The OCM is the key component in the CLG process. In recent years, metal oxides of Ni, Fe, Co, Cu, Mn, and Ce, $CaSO_4$, perovskite, etc. were investigated as OCs in CLC, chemical looping reforming, and chemical looping with oxygen uncoupling [33–41]. Several of these OCMs have also been considered for CLG of biomass as discussed below.

Huang et al. used natural hematite, that is, iron ore, as an OC in direct chemical looping conversion of biomass, for example, sawdust of pine, in a lab-scale fluidized bed reactor [42–44]. Batch experiments under argon atmosphere at 840°C focused on the investigation of the cyclic performance of OC. The presence of the iron ore evidently promoted the biomass conversion, acting as a gasifying medium similar to steam. The gas yield and carbon conversion increased from $0.75\,m_N^3/kg$ and 62.2% of biomass during pyrolysis experiments without iron ore as OC to $1.06\,m_N^3/kg$ and 87.6% in the presence of natural hematite. An optimum Fe_2O_3/C molar ratio of 0.23 was determined. The OC was gradually deactivated with reduction time increasing, inhibiting the carbon and hydrogen in biomass from being converted into synthesis gas. The fraction of Fe^{2+} increased from 0% to 47.1% after a reduction time of 45 min, which implied that active lattice oxygen of 49.8% was consumed. The components of the gas product in chemical looping mode (22% H_2, 12.3% CH_4, 50.6% CO, 11.6% CO_2, and 3.6% C_2H_y) were close to those in biomass pyrolysis (22.8% H_2, 12.9% CH_4, 53.2% CO, 6.6% CO_2, and 4.6% C_2H_y) as the cyclic number increased. The tar content was $10.3\,g/m_N^3$ in presence of iron ore and $36.2\,g/m_N^3$ without OC. The gas yield and carbon conversion decreased from $1.06\,m_N^3/kg$ and 87.6% in the 1st cycle to $0.93\,m_N^3/kg$ and 77.2% in the 20th cycle due to attrition and structural changes of OC. X-ray diffraction (XRD) analysis showed that the reduction extent of the OC increased with the cycles. Scanning electron microscope and pore structural analysis displayed that agglomeration was observed with the cycles. The Brunauer-Emmett-Teller (BET)-surface decreased from 1.05 to $0.63\,m^2/g$ while the average pore diameter increased from 20.3 to 43.7 nm. Further experiments were performed with the addition of steam. Comparing with biomass steam gasification, carbon conversion and gas yield increased by 7.5% and 11%, respectively, at a SBR of 0.85 at 800°C. The gas yield reached $1.41\,m_N^3/kg$ and the carbon conversion 93%. The tar content decreased by 51.5%. In this case, 62.3% of the lattice oxygen in the hematite particles was consumed in the biomass gasification. With the reaction temperature increasing from 750°C to 850°C, the gas yield increased from 1.12 to $1.53\,m_N^3/kg$, and carbon conversion increased from 77.2% to 95.5%. The gas concentration was gradually approaching that of steam gasification at the end stage of CLG since the active lattice oxygen was depleted with the proceeding reactions.

Ge et al. used natural hematite as an OC in a batch reactor and continuous $25\,kW_{th}$ dual reactor [45,46]. The continuous system consisted of a high velocity fluidized bed as an AR, a cyclone, a bubbling fluidized bed as a FR, and a loop-seal. Because of the intensive endothermic reactions in the FR, the gasification temperature decreased sharply for bed materials of 100 wt% silica sand.

Only when the hematite mass percentages reached 40 wt%, was the gasification temperature stable. Beside the acceleration of the process of biomass gasification and enhanced carbon conversion rate in the presence of hematite already observed by Huang et al. in their batch experiments, hematite could increase the heat-carrier capacity of the bed material in the continuous reactor. In the batch reactor, carbon conversion efficiency and fraction of CO and H_2 increased within the temperature range of 750–900°C. However, syngas yield in the continuous reactor reached the maximum of 0.64 m_N^3/kg at 850°C and a SBR of 1. In addition, the effect of hematite fraction on the gasification performance was similar between the batch reactor and the continuous reactor. When hematite fraction was above 40 wt%, a higher hematite fraction resulted in higher carbon conversion efficiency. However, a contrary trend was observed for the gas yield due to increased total oxidation of carbon to CO_2. Fe_3O_4 was the main reduced phase of hematite in the batch reactor, while FeO might exist in the continuous reactor.

Wei et al. used a synthetic Fe_2O_3/Al_2O_3 oxygen carrier with a mass ratio of $Fe_2O_3/Al_2O_3 = 7/3$ and pine sawdust as fuel in a 10 kW_{th} interconnected fluidized bed reactor [47]. The results indicated that the sawdust was partially oxidized to syngas by lattice oxygen from the OC. The syngas yield (0.95–1.17 m_N^3/kg), cold gas efficiency (41.6–61.1%), and carbon conversion (83.9–93.8%) increased with increasing operating temperature of 670–900°C. Also, the concentrations of CO, H_2, and CH_4 in the syngas increased at the elevated temperature, while the CO_2 fraction decreased. The feeding rate of biomass had a significant impact on the syngas composition and cold gas efficiency. They found an optimal value of feeding rate at 2.24 kg/h corresponding to the maximum cold gas efficiency in the tested reactor system of 70%. XRD analysis showed that the OC particles were reduced to Fe_3O_4. The surface area, total pore volume, and average pore size of the OC particles increased initially and then slightly decreased with the reaction proceeding. However, the OC particles could be well regenerated and maintained their crystalline structure after 60 h of operation without agglomeration.

To improve the performance of iron-based OCMs, the addition of NiO was investigated [48,49]. A NiO-modified iron ore OC was prepared by the impregnation method coupled with ultrasonic treatment [48]. The formation of spinel-type nickel iron oxide $NiFe_2O_4$ enhanced the reactivity of the OC and improved the reaction rate of char gasification. In TGA tests the reactivity of the OC increased with the increase of NiO loading. An optimal mass ratio of char/OC was determined at 4:6 with the aim of obtaining a maximum reaction rate. A relatively high carbon conversion of 55.6% was obtained. The OC was completely reduced into metallic iron and nickel, which can act as catalysts for char pyrolysis. Thus, conversion of biomass char was enhanced. Investigations in a 10 kW_{th} interconnected fluidized-bed reactor with Fe–Ni bimetallic oxides as OCs showed higher gasification efficiency (70.5%) compared to similar Fe_2O_3/Al_2O_3 oxygen carrier [49]. XRD results indicated that the OC was reduced to Fe_3O_4 (only) and Ni-Fe kamacite was formed during redox cycles improving the reactivity of the OC. Sintering of the OC was much less in comparison to the Ni-free OC.

Aghabararnejad et al. performed TGA measurements and kinetic modeling of Co, Mn, and Cu oxides for CLG [50]. Based on thermodynamic equilibrium, copper, manganese, and cobalt oxides have the highest oxygen release capacities among the different OCs. These OCs were deposited on alumina via incipient wetness impregnation. The weight loss of the $CuO–Cu_2O$ carrier, as measured in a TGA, was 10% while it was 7% for the $Co_3O_4–CoO$ and 3% for the $Mn_2O_3–Mn_3O_4$. The optimum operating temperature for the CuO oxygen carrier was 100°C higher compared to the other two at 950°C. The CuO carrier surface area decreased by 70%, while it was 30% and 60% with the Co_3O_4 and Mn_2O_3 carriers, respectively. Cobalt had a lower tendency to sinter at high temperature compared to either copper or manganese and showed a higher oxygen transport capacity and oxidation–reduction rates. Therefore, the authors suggested it as a potential OC despite its higher cost and toxicity.

In the literature, CLG is not only associated with the utilization of OCMs for supply of pure oxygen. CaO-based CLG aims at producing a hydrogen-rich gas by utilizing CaO for in situ CO_2 capture [51–53]. Fig. 6.3 shows the principle of this process. The system consists of the gasifier and a regenerator. In the gasifier, the biomass is gasified in the presence of steam to produce hydrogen-rich gas. CaO reacts with the generated CO_2 and forms $CaCO_3$. $CaCO_3$ is circulated to the regenerator to get calcined, releasing a highly concentrated stream of CO_2. The regenerated hot CaO is then cycled back to the gasifier. It also provides additional heat for the gasification process. In addition, the heat released by the exothermic carbonation reactions compensates the endothermic gasification reactions.

Acharya et al. [51] reported a system efficiency of 87.5% in the case of 100% CO_2 capture. A concentration of 71% H_2 and nearly 0% CO_2 in the producer gas was achieved in experiments performed in a batch-type fluidized-bed gasifier. Dual fluidized bed gasifiers, such as the Güssing gasifier, are well suited to combine gasification with sorption of carbon dioxide, as described in Section 5.2.1.

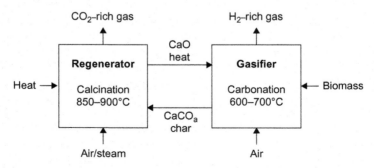

FIGURE 6.3 Schematic of Ca-based chemical looping gasification.

6.3 SUPERCRITICAL WATER GASIFICATION

Supercritical water gasification is a promising concept for converting wet biomass with high water content to syngas. Energy-consuming drying of the wet biomass is not required, resulting in higher energy efficiencies. Wet biomass with water content of more than 80 wt% can be used as feedstock. Even wastewater containing organic material can be used.

Above its critical point of $p = 22.12$ MPa and $T = 374.12°C$ (see Fig. 6.4), water achieves supercritical conditions. In the supercritical state, it cannot be distinguished between gas and liquid phase. Properties of the liquid and the gaseous phase are aligned at the critical point so that just one phase exists— the supercritical state. Supercritical water has the density of liquid water (0.3 g/cm^3 at the critical point) and the viscosity of water vapor (eg, $49·10^{-6}$ Pa s at 375°C and 22.5 MPa [54]). Moreover, supercritical water has a changed structure with separated polar water molecules corresponding to a decrease of the dielectric constant. With a lower dielectric constant, the solubility of organic polar substances increases and the solubility of inorganic ionic substances are reduced. Thus, supercritical water has unique properties as a solvent. The solubility of organic materials and gases is significantly increased and materials which are insoluble in water or water vapor can be dissolved, whereas the solubility of inorganic materials is decreased. For the gasification of biomass, the supercritical water is also an active reactant which results in a high hydrogen yield. Supercritical water gasification of biomass is typically performed at temperatures between 600°C and 650°C and at a pressure of about 30 MPa. Above 600°C, water acts as a strong oxidant. Carbon atoms are oxidized and CO_2 is preferably formed. Hydrogen atoms from water as well as from biomass are released to form hydrogen [55]. The main components in the produced gas are H_2, CH_4, and CO_2. CO content is typically low since CO reacts further by water

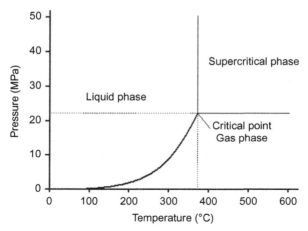

FIGURE 6.4 Schematic phase diagram of water.

gas shift and methanation reaction [56]. Tar and coke formation is inhibited by fast solution of the formed gas components in the supercritical water. At reaction temperatures below 450°C, CH_4 is the main component in the produced gas, whereas at reaction temperatures above 600°C hydrogen is dominant [57].

Gasification of organic hydrocarbons in supercritical water has already been investigated since the middle of the 1970s. However, broader interest in the research of supercritical water gasification of biomass has been gained in the last 10 years, expressed by a significantly increased number of publications in this research area.

To understand the influence of the reaction temperature, the pressure, the heating rate, and the concentration on the gasification process, many studies have been performed using model compounds, mostly glucose and cellulose (see eg, Refs. [58–61]). In addition, several studies on the usability of different biomass feedstocks have been performed in recent years, such as, for example, agricultural wastes, leather wastes, switchgrass, sewage sludge, algae, manure, olive mill wastewater, and black liquor (see eg, Refs. [56,62–67]).

The efficiency of supercritical water gasification can be improved by using catalysts in the process. The reaction temperature can be reduced and the reaction rate can be enhanced by catalysts. The investigation of catalytic gasification of biomass in supercritical water attracts a highly increasing interest. It is one of the newest research topics and the one with the highest growth rate in the research area of supercritical water science [68].

A decrease of the reaction temperature by using catalysts reduces the equipment as well as the operating costs. Furthermore, the conversion rate and the hydrogen yield can be enhanced. Several catalysts have been tested for supercritical water gasification of biomass including the typical reforming catalysts, such as Ni and Ru, Pt- and Rh-based catalysts, activated carbon, alkali metal-based materials, such as Trona, KOH, NaOH, K_2CO_3. Nickel-based catalysts are often investigated due to their relative low costs compared to noble metals and the high activity of Ni. Problems which occur with Ni catalysts are deactivation by carbon deposition and hydrothermal crystallite growth. Moreover, instability of the support material (Ni is mostly used supported on a metal oxide) under hydrothermal conditions can be a further problem. Noble metals, such as Ru, Rh, and Pt, show higher activity than Ni. In particular, Ru has a very high activity and a high resistance against oxidation and sintering under hydrothermal conditions. A problem of noble metal catalysts is deactivation by sulfur which requires an efficient preremoval of sulfur from the feed. The catalytic activity of activated carbon in supercritical water gasification of biomass has been investigated by several researchers (see eg, Refs. [69,70]). Activated carbon has a high stability in supercritical water—its gasification rate in supercritical water is very low. It is a very cheap catalyst material and shows a high activity for biomass gasification in supercritical water at high temperatures above 600°C. An alternative to heterogeneous catalysts are homogeneous catalysts which are mixed and dissolved in the feed. Alkaline metal catalysts, such as KOH, NaOH,

K_2CO_3, $Ca(OH)_2$, and $KHCO_3$, have gained interest as homogeneous catalysts for supercritical water gasification of biomass. Alkaline metals catalyzes the water–gas shift reaction which results in a significant increase of the hydrogen yield and reduces CO. Attention should be paid to the low solubility of inorganic salts in supercritical water. Combining heterogeneous metal catalysts and homogeneous alkaline metal catalysts showed enhanced efficiency in the gasification of biomass in supercritical water. For example, this has been shown for the combination of Raney-Ni and NaOH [71] and for Ru supported on alumina combined with NAOH and $Ca(OH)_2$ [72]. For more detailed information on catalytic supercritical water gasification, the reader is referred to the review papers of Elliott [73], Guo et al. [57], and Azadi and Farnood [74].

Reactor design and engineering of the process is a very important issue which has to take into account several requirements. Batch reactors are not an option for commercial application, only continuous flow reactors offer an economic production and industrial scale-up. The general flow scheme of a continuous process is shown in Fig. 6.5. The main unit operations of the process are heat exchange, reaction, and liquid/gas separation.

There are high requirements on the materials used in the system. Special corrosion and high pressure and high temperature resistant materials have to be used. This results in high investment costs.

Heat recovery from the reactor effluent is required to heat up the high amount of water in the feed efficiently. Without this heat recovery by using a heat exchanger, the overall energy efficiency of the process would be very low and the energy to heat up the feed to the reaction temperature of, for example, 600°C could even be higher than the energy content of the biomass if the water content is higher than 80% [75]. Besides heat recovery from the reactor effluent, heating rate of the feed is very important for supercritical water gasification. A low heating rate of the feed results in coke and tar formation and decreases the gas yield. A challenge for an industrial scale-up of a continuous supercritical water gasification process is precipitation of salts contained in the biomass which can lead to plugging problems. The energy efficiency of the supercritical water gasification of biomass has been studied by different researchers. Marias et al. [76] modeled the gasification of vinasse in supercritical water. They showed that the best gasification efficiency of their process of about 87% was achieved at a reaction temperature of 600°C. By thermodynamic modeling of a supercritical

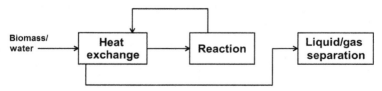

FIGURE 6.5   General flow scheme and main unit operations of a continuous supercritical water biomass gasification process.

water gasification process for hydrogen production from wet biomass, Lu et al. [77] showed an energy and exergy efficiency of their process of more than 40%. Energy loss mainly results from heat transfer in the heat exchanger, the cooler, the preheater, and the reactor. High heat transfer efficiency increases noticeably the energy efficiency of the whole system. As an alternative to very efficient heat exchangers, oxygen can be added to the process in a small amount to make the process energetically self-sustainable with only a small loss of the heating value of the produced syngas [78]. A polygeneration process for the production of methanol and power by supercritical water gasification of glycerol has been studied by Gutierrez Ortiz et al. [79]. Power was produced by a turbine and a fuel cell in their study. They calculated an overall net energy efficiency of their proposed process of 38% for a gasification temperature of 1000°C.

Even though progress has been made in supercritical water gasification of biomass in recent years and this technology seems to be very interesting especially for wet biomass, technical solutions for large-scale commercial installations still need to be developed.

REFERENCES

[1] Fu Q, Huang Y, Niu M, Yang G, Shao Z. Experimental and predicted approaches for biomass gasification with enriched air-steam in a fluidised bed. Waste Manage Res 2014;32:988–96.

[2] Pu G, Zhou HP, Hao GT. Study on pine biomass air and oxygen/steam gasification in the fixed bed gasifier. Int J Hydrogen Energy 2013;38:15757–63.

[3] Silva V, Couto N, Alexandre JL, Rouboa A. Analysis of the pine biomass gasification to obtain H_2 enriched syngas. Adv Sci Lett 2013;19:946–9.

[4] Huynh CV, Kong SC. Performance characteristics of a pilot-scale biomass gasifier using oxygen-enriched air and steam. Fuel 2013;103:987–96.

[5] Barisano D, Canneto G, Nanna F, Alvino E, Pinto G, Villone A, et al. Steam/oxygen biomass gasification at pilot scale in an internally circulating bubbling fluidized bed reactor. Fuel Process Technol 2016;141:74–81.

[6] Broer KM, Woolcock PJ, Johnston PA, Brown RC. Steam/oxygen gasification system for the production of clean syngas from switchgrass. Fuel 2015;140:282–92.

[7] Sandeep K, Dasappa S. Oxy-steam gasification of biomass for hydrogen rich syngas production using downdraft reactor configuration. Int J Energy Res 2014;38:174–88.

[8] Pinto F, André RN, Franco C, Carolino C, Costa R, Miranda M, et al. Comparison of a pilot scale gasification installation performance when air or oxygen is used as gasification medium 1. Tars and gaseous hydrocarbons formation. Fuel 2012;101:102–14.

[9] Gil J, Aznar MP, Caballero MA, Frances E, Corella J. Biomass gasification in fluidized bed at pilot scale with steam-oxygen mixtures. Product distribution for very different operating conditions. Energy Fuels 1997;11:1109–18.

[10] Perez P, Aznar PM, Caballero MA, Gil J, Martin JA, Corella J. Hot gas cleaning and upgrading with a calcined dolomite located downstream a biomass fluidized bed gasifier operating with steam-oxygen mixtures. Energy Fuels 1997;11:1194–203.

[11] Caballero MA, Aznar MP, Gil J, Martin JA, Frances E, Corella J. Commercial steam reforming catalysts to improve biomass gasification with steam-oxygen mixtures. 1. Hot gas upgrading by the catalytic reactor. Ind Eng Chem Res 1997;36:5227–39.

[12] Simeone E, Siedlecki M, Nacken M, Heidenreich S, de Jong W. High temperature gas filtra-
      tion with ceramic candles and ashes characterisation during steam-oxygen blown gasification
      of biomass. Fuel 2013;108:99–111.
[13] Meng X, Benito P, de Jong W, Basile F, Verkooijen AHM, Fornasari G, et al. Steam–O2
      blown circulating fluidized-bed (CFB) biomass gasification: characterization of different
      residual chars and comparison of their gasification behavior to thermogravimetric (TG)-
      derived pyrolysis chars. Energy Fuels 2012;26:722–39.
[14] Siedlecki M, de Jong W. Biomass gasification as the first hot step in clean syngas production
      process—gas quality optimization and primary tar reduction measures in a 100 kW thermal
      input steam-oxygen blown CFB gasifier. Biomass Bioenergy 2011;35:S40–62.
[15] Meng X, de Jong W, Fu N, Verkooijen AHM. Biomass gasification in a $100 kW_{th}$ steam-
      oxygen blown circulating fluidized bed gasifier: effects of operational conditions on product
      gas distribution and tar formation. Biomass Bioenergy 2011;35:2910–24.
[16] Siedlecki M, Nieuwstraten R, Simeone E, de Jong W, Verkooijen AHM. Effect of magnesite
      as bed material in a $100 kW_{th}$ steam-oxygen blown circulating fluidized-bed biomass gasifier
      on gas composition and tar formation. Energy Fuels 2009;23:5643–54.
[17] Tan X, Wang Z, Meng B, Meng X, Li K. Pilot-scale production of oxygen from air using
      perovskite hollow fibre membranes. J Memb Sci 2010;352:189–96.
[18] Xu SJ, Thomson WJ. Oxygen permeation rates through ion-conducting perovskite mem-
      branes. Chem Eng Sci 1999;54:3839–50.
[19] Kneer R, Toporov D, Forster M, Christ D, Broeckmann C, Pfaff E, et al. OXYCOAL-AC:
      towards an integrated coal-fired power plant process with ion transport membrane-based oxy-
      gen supply. Energy Environ Sci 2010;3:198–207.
[20] Eichhorn Colombo K, Kharton VV, Bolland O. Simulation of an oxygen membrane-based gas
      turbine power plant: dynamic regimes with operational and material constraints. Energy Fuels
      2010;24:590–608.
[21] Castillo R. Thermodynamic analysis of a hard coal oxyfuel power plant with high tempera-
      ture three-end membrane for air separation. Appl Energy 2011;88:1480–93.
[22] Kunze C, De S, Spliethoff H. A novel IGCC plant with membrane oxygen separation and
      carbon capture by carbonation–calcinations loop. Int J Greenhouse Gas Control 2011;5:
      1176–83.
[23] Gupta S, Adams JJ, Wilson JR, Eddings EG, Mahapatra MK, Singh P. Performance and
      post-test characterization of an OTM system in an experimental coal gasifier. Appl Energy
      2016;165:72–80.
[24] Puig-Arnavat M, Soprani S, Søgaard M, Engelbrecht K, Ahrenfeldt J, Henriksena UB, et al.
      Integration of mixed conducting membranes in an oxygen–steam biomass gasification pro-
      cess. RSC Adv 2013;3:20843–54.
[25] Antonini T, Gallucci K, Foscolo PU. A biomass gasifier including an ionic transport mem-
      brane system for oxygen transfer. Chem Eng Trans 2014;37:91–6.
[26] Antonini T, Gallucci K, Anzoletti V, Stendardo S, Foscolo PU. Oxygen transport by ionic
      membranes: correlation of permeation data and prediction of char burning in a membrane-
      assisted biomass gasification process. Chem Eng Process 2015;94:39–52.
[27] He F, Galinsky N, Li FX. Chemical looping gasification of solid fuels using bimetallic oxy-
      gen carrier particles e feasibility assessment and process simulations. Int J Hydrogen Energy
      2013;38:7839–54.
[28] Aghabararnejad M, Gregory SP, Jamal C. Techno-economic comparison of a $7\text{-}MW_{th}$ bio-
      mass chemical looping gasification unit with conventional systems. Chem Eng Technol
      2015;38:876–8.

[29] Ryden M, Lyngfelt A, Mattisson T. Chemical-looping combustion and chemical-looping reforming in a circulating fluidized-bed reactor using Ni-based oxygen carriers. Energy Fuel 2008;22:2585–97.

[30] Li FX, Kim HR, Sridhar D, Wang F, Zeng L, Chen J, et al. Syngas chemical looping gasification process: oxygen carrier particle selection and performance. Energy Fuel 2009;23: 4182–9.

[31] Xie QL, Borges FC, Cheng YL, Wan YQ, Li Y, Lin XY, et al. Fast microwave-assisted catalytic gasification of biomass for syngas production and tar removal. Bioresour Technol 2014;156:291–6.

[32] Nordgreen T, Nemanova V, Engvall K, Sjöström K. Iron-based materials as tar depletion catalysts in biomass gasification: dependency on oxygen potential. Fuel 2012;95:71–8.

[33] Cao Y, Pan WP. Investigation of chemical looping combustion by solid fuels. 1. Process analysis. Energy Fuel 2006;20:1836–44.

[34] Lea-Langton A, Zin RM, Dupont V, Twigg MV. Biomass pyrolysis oils for hydrogen production using chemical looping reforming. Int J Hydrogen Energy 2012;37:2037–43.

[35] Ryden M, Leion H, Mattisson T, Lyngfelt A. Combined oxides as oxygen-carrier material for chemical-looping with oxygen uncoupling. Appl Energy 2014;113:1924–32.

[36] Niu X, Shen LH, Gu HM, Song T, Xiao J. Sewage sludge combustion in a CLC process using nickel-based oxygen carrier. Chem Eng J 2015;260:631–41.

[37] Song T, Zheng M, Shen LH, Zhang T, Niu X, Xiao J. Mechanism investigation of enhancing reaction performance with $CaSO_4/Fe_2O_3$ oxygen carrier in chemical looping combustion of coal. Ind Eng Chem Res 2013;52:4059–71.

[38] Pans MA, Gayan P, Abad A, Garcia-Labiano F, de Diego LF, Adanez J. Use of chemically and physically mixed iron and nickel oxides as oxygen carriers for gas combustion in a CLC process. Fuel Process Technol 2013;115:152–63.

[39] Mattisson T, Jerndal E, Linderholm C, Lyngfelt A. Reactivity of a spray-dried $NiO/NiAl_2O_4$ oxygen carrier for chemical looping combustion. Chem Eng Sci 2011;66:4636–44.

[40] Arjmand M, Azad A, Leion H, Lyngfelt A, Mattisson T. Prospects of Al_2O_3 and $MgAl_2O_4$-supported CuO oxygen carriers in chemical-looping combustion (CLC) and chemical-looping with oxygen uncoupling (CLOU). Energy Fuel 2011;25:5493–502.

[41] Shulman A, Cleverstam E, Mattisson T, Lyngfelt A. Manganese/iron, manganese/nickel, and manganese/silicon oxides used in chemical-looping with oxygen uncoupling (CLOU) for combustion of methane. Energy Fuel 2009;23:5269–75.

[42] Huang Z, He F, Feng Y, Zhao K, Zheng A, Chang S, et al. Synthesis gas production through biomass direct chemical looping conversion with natural hematite as an oxygen carrier. Bioresour Technol 2013;140:138–45.

[43] Huang Z, He F, Feng Y, Liu R. Characteristics of biomass gasification using chemical looping with iron ore as an oxygen carrier. Int J Hydrogen Energy 2013;38:14568–75.

[44] Huang Z, He F, Zheng A. Synthesis gas production from biomass gasification using steam coupling with natural hematite as oxygen carrier. Energy 2013;53:244–51.

[45] Ge H, Guo W, Shen L, Song T, Xiao J. Experimental investigation on biomass gasification using chemical looping in a batch reactor and a continuous dual reactor. Chem Eng J 2016;286:689–700.

[46] Ge H, Guo W, Shen L, Song T, Xiao J. Biomass gasification using chemical looping in a $25\,kW_{th}$ reactor with natural hematite as oxygen carrier. Chem Eng J 2016;286:174–83.

[47] Wei G, He F, Huang Z, Zheng A, Zhao K, Li H. Continuous operation of a $10\,kW_{th}$ chemical looping integrated fluidized bed reactor for gasifying biomass using an iron-based oxygen carrier. Energy Fuel 2015;29:233–41.

[48] Huang Z, He F, Feng Y, Zhao K, Zheng A, Chang S, et al. Biomass char direct chemical looping gasification using NiO-modified iron ore as an oxygen carrier. Energy Fuel 2014;28:183–91.

[49] Wie G, He F, Zhao Z, Huang Z, Zheng A, Zhao K, et al. Performance of Fe-Ni bimetallic oxygen carriers for chemical looping gasification of biomass in a $10kW_{th}$ interconnected circulating fluidized bed reactor. Int J Hydrogen Energy 2015;40:16021–32.

[50] Aghabararnejad M, Patience GS, Chaouki J. TGA and kinetic modelling of Co, Mn and Cu oxides for chemical looping gasification (CLG). Can J Chem Eng 2014;92:1903–10.

[51] Acharya B, Dutta A, Basu P. Chemical looping gasification of biomass for hydrogen-enriched gas production with in-process carbon dioxide capture. Energy Fuel 2009;23:5077–83.

[52] Udomsirichakorn J, Basu P, Salama PA, Acharya B. CaO-based chemical looping gasification of biomass for hydrogen-enriched gas production with in situ CO_2 capture and tar reduction. Fuel Process Technol 2014;127:7–12.

[53] Di Felice L, Courson C, Jand N, Gallucci K, Foscolo PU, Kiennemann A. Catalytic biomass gasification: simultaneous hydrocarbons steam reforming and CO_2 capture in a fluidised bed reactor. Chem Eng J 2009;154:375–83.

[54] Grigull U, Mayinger F, Bach J. Viskosität, Wärmeleitfähigkeit und Prandtl-Zahl von Wasser und Wasserdampf. Wärme Stoffübertragung 1968;1:15–34.

[55] Feng W, van der Kooi HJ, de Swaan Arons J. Biomass conversion in subcritical and supercritical water: driving force, phase equilibria, and thermodynamic analysis. Chem Eng Process 2004;43:1459–67.

[56] Yanik J, Ebale S, Kruse A, Saglam M, Yüksel M. Biomass gasification in supercritical water: part1. Effect of the nature of biomass. Fuel 2007;86:2410–5.

[57] Guo Y, Wang SZ, Xu DH, Gong YM, Ma HH, Tang XY. Review of catalytic supercritical water gasification for hydrogen production from biomass. Renewable Sustainable Energy Rev 2010;14:334–43.

[58] Hendry D, Venkitasamy C, Williams PT, Jacoby W. Exploration of the effect of process variables on the production of high-value fuel gas from glucose via supercritical water gasification. Bioresour Technol 2011;102:3480–7.

[59] Promdej C, Matsumura Y. Temperature effect on hydrothermal decomposition of glucose in sub- and supercritical water. Ind Eng Chem Res 2011;50:8492–7.

[60] Susanti RF, Dianningrum LW, Yum T, Kim Y, Lee BG, Kim J. High-yield hydrogen production from glucose by supercritical water gasification without added catalyst. Int J Hydrogen Energy 2012;37:11677–90.

[61] Caputo G, Rubio P, Scargiali F, Marotta G, Brucato A. Experimental and fluid dynamic study of continuous supercritical water gasification of glucose. J Supercrit Fluids 2016;107:450–61.

[62] Byrd AJ, Kumar S, Kong L, Ramsurn H, Gupta RB. Hydrogen production from catalytic gasification of switchgrass biocrude in supercritical water. Int J Hydrogen Energy 2011;36:3426–33.

[63] Chen Y, Guo L, Cao W, Jin H, Guo S, Zhang X. Hydrogen production by sewage sludge gasification in supercritical water with a fluidized bed reactor. Int J Hydrogen Energy 2013;38:12991–9.

[64] Miller A, Hendry D, Wilkinson N, Venkitasamy C, Jacoby W. Exploration of the gasification of spirulina algae in supercritical water. Bioresour Technol 2012;119:41–7.

[65] Yakaboylu O, Harinck J, Smit KGG, de Jong W. Supercritical water gasification od manure: a thermodynamic equilibrium modeling approach. Biomass Bioenergy 2013;59:253–63.

[66] Kıpcak E, Akgün M. Oxidative gasification of olive mill wastewater as a biomass source in supercritical water: effects on gasification yield and biofuel composition. J Supercrit Fluids 2012;69:57–63.

[67] Sricharoenchaikul V. Assessment of black liquor gasification in supercritical water. Bioresour Technol 2009;100:638–43.

[68] Kamler J, Soria JA. Supercritical water gasification of municipal sludge: a novel approach to waste treatment and energy recovery; 2012. <http://dx.doi.org/10.5772/51048>.

[69] Antal MJ, Allen SG, Schulman D, Xu X, Divilio RJ. Biomass gasification in supercritical water. Ind Eng Chem Res 2000;39:4040–53.

[70] Nakamura A, Kiyonaga E, Yamamura Y, Shimizu Y, Minowa T, Noda Y, et al. Gasification of catalyst-suspended chicken manure in supercritical water. J Chem Eng Jpn 2008;41:433–40.

[71] Jin H, Lu Y, Guo L, Zhang X, Pei A. Hydrogen production by supercritical water gasification of biomass with homogeneous and heterogeneous catalyst. Adv Condens Matter Phys 2014 Article ID 160565, 9 pp. <http://dx.doi.org./10.1155/2014/160565>.

[72] Onwudili JA, Williams PT. Hydrogen and methane selectivity during alkaline super-critical water gasification of biomass with ruthenium-alumina catalyst. Appl Catal B 2013;132–133:70–9.

[73] Elliott DC. Catalytic hydrothermal gasification of biomass. Biofuels Bioprod Biorefin 2008;2:254–65.

[74] Azadi P, Farnood R. Review of heterogeneous catalysts for sub- and supercritical water gasification of biomass and wastes. Int J Hydrogen Energy 2011;36:9529–41.

[75] Kruse A. Hydrothermal biomass gasification. J Supercrit Fluids 2009;47:391–9.

[76] Marias F, Letellier S, Cezac P, Serin JP. Energetic analysis of gasification of aqueous biomass in supercritical water. Biomass Bioenergy 2011;35:59–73.

[77] Lu Y, Guo L, Zhang X, Yan Q. Thermodynamic modeling and analysis of biomass gasification for hydrogen production in supercritical water. Chem Eng J 2007;131:233–44.

[78] Castello D, Fiori L. Supercritical water gasification of biomass: thermodynamic constraints. Bioresour Technol 2011;102:7574–82.

[79] Gutierrez Ortiz FJ, Serrera A, Galera S, Ollero P. Methanol synthesis from syngas obtained by supercritical water reforming of glycerol. Fuel 2013;105:739–51.

Chapter 7

Polygeneration Strategies

7.1 COMBINED HEAT AND POWER PRODUCTION

Power production from fuels is generally coupled with the production of heat. If the heat can be used, the overall process efficiency can be significantly increased. In contrast to electricity, which can be easily transported and distributed, heat has to be produced close to the user. Due to this fact, it is preferred to have decentralized heat and power production by smaller units than to have large units of several hundreds of MW as typically used for power production. Examples of small decentralized combined heat and power production (CHP) units where heat can beneficially be used are: (1) CHP plants of a few MW coupled to district heating systems; (2) CHP plants below 1 MW for heating large public or commercial buildings, public swimming pools, hospitals, hotels, or apartment buildings; and (3) CHP plants installed at pulp and paper plants, at saw mills, etc., where biomass waste can directly be used to produce electricity, heat, or steam for the local plant. CHP by biomass combustion is already widely used [1]. However, CHP by biomass gasification compared to combustion offers some advantages: (1) higher biomass to power efficiency [2,3]; (2) higher flexibility concerning the used biomass feedstock; and (3) higher flexibility concerning the applied power generation process [4]. First CHP applications by biomass gasification started at the beginning of the 1990s [5]. CHP biomass gasification plants with capacities of a few MW of biomass input have demonstrated several years of successful operation showing the overall reliability of the technology and the achievable high process efficiency. Some examples are the 5.2 MW_{th} Harboøre plant (start-up in 1993) [2,6,7] and the 26 MW_{th} Skive plant (start-up in 2006) [7] in Denmark, the 8 MW_{th} plant in Güssing (start-up in 2002) [8], the 2 MW_{th} plant in Wiener Neustadt (in operation from 2003 to 2007) [9,10], the 8.5 MW_{th} plant in Oberwart (start-up in 2008) [7], and the 15 MW_{th} plant in Villach (in operation from 2010 to 2013) in Austria, as well as the 15 MW_{th} plant in Ulm (start-up in 2012) in Germany [11]. In almost each of these plants, overall process efficiencies of about 90% and biomass to electricity efficiencies from 25% to 31% are reported. In all cases gas engines are used for electricity generation showing the increased efficiency of newer gas engine types. The heat produced is used for district heating.

Advanced Biomass Gasification. DOI: http://dx.doi.org/10.1016/B978-0-12-804296-0.00007-5

A way to increase the amount of generated electricity is to use additionally an organic Rankine cycle (ORC). In the $15\,MW_{th}$ plant in Ulm in Germany, an ORC is installed in addition to two gas engines for electricity production [11]. By using ORC, some of the heat (10–15% of the heat) [7] can be additionally converted into electricity. By combining gas engines with ORC, biomass to electricity efficiencies of more than 40% are achievable.

The integrated gasification combined cycle (IGCC) process is another approach to increase the electricity efficiency. In an IGCC process, a gas turbine and a steam turbine are combined to generate electricity. The IGCC process has been used for coal gasification since the mid-1990s, showing electricity efficiencies of up to 46% for plant sizes of 200–300 MW [12,13]. Theoretical investigations (see eg, Ref. [14]) show that even overall electrical efficiencies of up to 53% could be possible with an IGCC process. For biomass, the IGCC process has been applied in the $18\,MW_{th}$ ($6\,MW_{el}$ and $9\,MW_{th}$ for district heating) demonstration plant in Värnamo in Sweden [15,16]. The plant was operated from 1993 to 1999 when the demonstration program was completed. The plant had a net electrical efficiency of 32% and a total net efficiency of 83% [15]. Since small steam turbines have a low electrical efficiency [17], an IGCC process is only interesting for larger scales.

A technology for small power generation units which has already gained a high interest is the fuel cell technology. Fuel cells offer the advantages of very high electrical efficiencies, an environmental friendly operation and they can be used from some hundreds of kilowatts down to 1 kilowatt for very small scale applications. For the combination with biomass gasification, Solid Oxide Fuel Cells (SOFC) are considered as a good option since they operate at high temperatures with hydrogen, carbon monoxide, and methane and their mixtures as fuel [18]. The high exhaust temperature of the SOFC has an advantage of using the heat for CHP application. However, even if the biomass gasification syngas composition is generally well suitable as fuel for a SOFC, it has to be cleaned to achieve the high requirements of the fuel cell with regard to impurities. A comprehensive literature review of the different contaminants in the raw biomass syngas, their influence on the performance of a SOFC, and techniques to remove them was presented by Aravind and de Jong [18]. They concluded that it is possible to clean the syngas to meet the requirements of a SOFC. They proposed a gas cleaning set-up comprising a series of fixed bed reactors and two ceramic hot gas filters.

Many performance models and theoretical analyses of the combination of SOFCs and biomass gasifiers show the high potential of this combination with electrical efficiencies from 34% to 45% and CHP efficiencies from 74% to 80% [19–23].

By combining a SOFC with an additional Micro Gas Turbine (MGT), a further increase of the electrical efficiency is possible [24,25]. The excess fuel in the exhaust gas from the SOFC is used by the MGT to generate additional electricity. The combination of a SOFC and a MGT with a biomass gasifier offers a very efficient power production solution for small decentralized CHP plants. Recent modeling studies of Bang-Møller et al. [26] showed an electrical efficiency of 58% and a CHP efficiency of 87.5% for an optimized process using the two stage Viking gasifier combined with a SOFC and a MGT. Morandin et al. [27] have recently modeled nine different system configurations using either an internally circulating fluidized bed gasifier or the Viking two stage gasifier and different combinations as CHP unit, just a SOFC, a SOFC combined with a steam cycle, and a SOFC combined with a MGT. Their thermo-economic analysis showed that the combination of a fast internally circulating fluidized bed gasifier with a pressurized operated SOFC and MGT is the most promising configuration with an electrical efficiency of 65%.

However, the combination of SOFC and biomass gasification has not been demonstrated on full scale and for long-term operation yet. Experimental investigations have only been performed with single cells or very small stacks, and for short test durations of about 5–150 h (see eg, Refs. [28–32]). Only one test has been performed for longer test duration of 1200 h [33].

The following conclusions can be drawn for the operation of SOFC with produced gas from biomass gasification based on published experimental results:

1. SOFCs can successfully be operated by using cleaned biomass gasification syngas. The general feasibility of the combination of biomass gasification with a SOFC is shown. However, demonstration on full scale and for long-term operation is still missing.
2. The cell performance depends on the syngas composition and fluctuations of both can be correlated [29].
3. The risk of carbon deposition increases with higher concentration of methane and tars. However, carbon deposition can be prevented by addition of steam.
4. On the other hand, if steam content is too high, the risk of nickel oxidation of the anode is increased resulting in cell performance degradation [32].
5. Inefficient particle removal from the syngas causes ash deposits and prevents a smooth long-term operation [28]. For this reason fly ash and char particles have to be removed efficiently from the syngas.
6. Besides particle removal, efficient removal of gaseous impurities, such as tars, sulfur compounds, alkalis, chlorides, and ammonia, is important to prevent poisoning of the anodes [18].
7. Attention has to be paid to the syngas composition. Higher concentrations of methane, propane, and higher hydrocarbons in the biomass syngas increase the rate of internal reforming, if no separate external reformer unit is used, and cause thermal stress, which may result in cracking of the cells [29].

Finally, the question of which one of the available technologies for CHP should be favored cannot be answered. Which CHP technology is preferred depends on specific conditions on a case by case basis. One important condition beside investment costs, earnings or incentives for heat and power supply, and availability and cost of biomass is the local annual heat demand. Gas engines are well proven and have a relatively high electrical efficiency at relatively low investment costs. SOFCs and the combination of SOFCs with MGTs are promising technologies which offer very high power generation efficiencies even at very small size. However, long-term testing on full scale size is still missing and will be an important topic for future investigations.

7.2 SYNTHETIC NATURAL GAS, HEAT AND POWER PRODUCTION

Synthetic natural gas (SNG) from biomass is considered as a renewable clean fuel substitute for fossil fuels in heating, CHP, and transportation systems. Bio-SNG is also regarded as a flexible renewable energy carrier. The existing natural gas pipeline system offers the advantage of an easy transportation and storage of the SNG. Gas heaters, boilers, and engines as well as natural gas cars and buses are available as established clean technologies for the use of SNG. Several methanation processes were developed in the past for the production of SNG from coal gasification syngas. One of these processes is, for example, the well-known Lurgi process using a series of adiabatic fixed bed reactors. A comprehensive review on methanation technologies has been given by Kopyscinski et al. [34].

Before the methanation reaction can be performed, the produced syngas has to be cleaned and conditioned. The gas cleaning aims to remove particulates, tars, alkali, and sulfur components. The removal of these components has to be very efficient in order to protect downstream catalysts from poisoning. Conditioning of the gas comprises typically a water gas shift to adjust the needed H_2/CO ratio to 3 or more for the methanation reaction. After the methanation, water and CO_2 as well as unreacted hydrogen and impurities of ammonia are removed from the SNG to achieve the required gas quality, for example, to deliver the SNG to the gas grid. More details of the process are given in the literature (see eg, Refs. [34–36]).

In the last 10 years, the production of SNG from biomass gasification syngas has gained increasing interest and has been investigated by some research groups, such as the Energy Research Center of the Netherlands (ECN) and the Paul Scherrer Institute (PSI) in Switzerland [34].

Recently, the feasibility to produce SNG from biomass gasification syngas has been demonstrated in a 1 MW scale at the Güssing plant [37]. ECN operates a $0.8\,MW_{th}$ SNG pilot unit [38]. A thorough investigation by Wirth and Markard showed that Bio-SNG plants require a size of 20 MW or larger to be economical since the process including gasification, gas cleaning, catalytic methanation, and CO_2 separation is complex and expensive [39].

The first commercial Bio-SNG plant of a size of 100 MW will be built in the GoBiGas project in Göteborg in Sweden [40,41]. In the first phase of the project, a 20 MW plant has been erected. The GoBiGas facility was inaugurated on Mar. 12, 2014. The facility converts waste wood to SNG via gasification, followed by gas cleaning and methane production. In Dec. 2014, methane produced by GoBiGas was injected into the natural gas grid [42]. It is planned to build an 80 MW plant in the second phase of the project. There are further plans for large commercial Bio-SNG plants, for example, the 200 MW Bio2G project of EOn in Sweden [43].

Finally, there is the question of whether the effort to produce SNG from the syngas is valuable. If the SNG is used for domestic heating and cooking, it would be a shorter chain to use the syngas directly instead of having an additional production step to produce SNG and to increase the costs and drop the process efficiency. The reason why the additional step to produce SNG is taken is that the existing distribution system for natural gas can easily be used for SNG which is a big advantage and justifies the higher effort. In areas where no natural gas distribution system is available it would be more beneficial to use the cleaned syngas directly for household heating and cooking as has already been practised, for example, in China in rural areas [44]. However, a risk of the direct use of the syngas is the toxicity of the CO contained in the gas.

Compared to domestic heating by wood combustion, Bio-SNG is less efficient and about twice the wood quantity would be required to produce the same quantity of heat [45]. Also power production by using Bio-SNG is less efficient than using the syngas from biomass gasification directly.

However, Bio-SNG as renewable fuel for transportation can have an efficiency advantage. Felder and Dones [46] showed in their evaluation of the ecological impact of Bio-SNG that the preferential use of SNG is as a transport fuel for cars to substitute oil-based fuels. Fahlen and Ahlgren [47] concluded in their study that it is more economical to use Bio-SNG as a vehicle fuel than it is to use it for power and heat production. Ahman [48] showed in his assessment of Bio-SNG as a transport fuel that Bio-SNG can be produced cost-efficiently already at low to medium scale plants of 20–100 MW, whereas the production of liquid biofuels requires larger scale plants to be competitive. Additionally, he stated that the production efficiency of Bio-SNG is higher than for DME, methanol, or Fischer–Tropsch (FT) diesel.

Production of Bio-SNG in a polygeneration plant offers several process possibilities with high overall process efficiencies of up to 90%. For example, excess heat of the process can be used for district heating [47] or part of the excess heat can be transformed in a Rankine cycle to coproduce power [49]. The integration of SNG and electricity via biomass gasification in a district heating system offers economic benefit and reduces CO_2 emission [50]. Polygeneration of Bio-SNG can offer a high flexibility on market demands by adjusting the requested amounts of Bio-SNG, heat and power or just balancing the demands by producing the Bio-SNG as energy storage.

In a polygeneration system there is also the possibility to operate the gasifier with the aim to produce a syngas with a higher content of methane and to separate this methane directly from the syngas avoiding the need of a subsequent methanation process. The remaining syngas can then be used for heat and power generation, for example.

7.3 BIOFUELS, HEAT AND POWER PRODUCTION

Comparable to Bio-SNG, biofuels, such as FT diesel, DME, and methanol, are considered as renewable clean fuel substitutes for fossil transportation or heating fuels. Additionally, liquid biofuels generate less emission than oil-based fuels.

Since the use of liquid fuels for cars and trucks is state-of-the-art, and accordingly an expanded infrastructure of oil-based refueling stations exists globally, liquid biofuels can be relatively easily introduced and used in the market. This is the main advantage for the use of liquid biofuels for transportation compared to SNG or hydrogen from biomass.

Polygeneration of liquid biofuels, heat and power offers as the main advantages high process flexibility and process efficiencies up to about 90% [2] which is similar to the polygeneration of Bio-SNG, heat and power described in Section 7.2. The advantage of a polygeneration versus a stand-alone production has been concluded from several techno-economic analyses.

Narvaez et al. [51] showed in a recent case study that a polygeneration plant for the production of methanol and power has a better performance and higher flexibility compared to the separate stand-alone production plants. Furthermore, they showed savings of the syngas consumption as well as the possibility to compensate a decrease of the catalyst activity by an increase of the syngas feed rate for the methanol production route and using the unreacted syngas afterwards for power production.

Haro et al. [52] concluded in their assessment of 12 different process concepts based on DME as intermediate and considering ethanol, methyl acetate, DME, H_2, and electricity as final products that the highest internal rate of return is given by a concept of producing methyl acetate, DME, and electricity and that the polygeneration is more profitable than a single product plant.

Meerman et al. [53] concluded in their study that the economics of a flexible polygeneration FT-liquid facility is better than its stand-alone counterpart.

Djuric Ilic et al. [54] showed in their study that polygeneration of biofuels with coproduction of heat and power has a higher reduction of greenhouse gas emissions than the stand-alone production of biofuels.

Comparing the production costs of biofuels to fossil fuels, different techno-economic analyses have shown that biofuels have to be supported by tax or CO_2 incentives in order to be competitive (see eg, Refs. [55,56]). The production costs decrease with the increasing size of the production plant [57].

7.4 HYDROGEN AND HEAT PRODUCTION

Hydrogen generated from biomass can be an interesting, environmentally friendly, and renewable energy carrier mainly for the use in fuel cells in stationary as well as transportation applications. Gasification of biomass produces a syngas rich in hydrogen if steam or oxygen and steam are used as gasification agent. If air is used as gasification medium, the nitrogen dilutes the syngas leading to a low hydrogen concentration and a high effort of hydrogen purification.

To separate and purify the hydrogen from the syngas pressure swing adsorption or membranes can be used. To achieve high hydrogen content in the syngas a reforming stage and additionally a water gas shift stage typically follows the gasifier.

Toonssen et al. [58] modeled 10 process configurations with and without heat recovery based on different gasifiers. They calculated exergy efficiencies between 45% and 50% for the cases of hydrogen production without heat recovery and between 62% and 66% with heat recovery. The hydrogen yield for the different gasifiers was between 97 and 106 g/kg of dry biomass. Bhattacharya et al. [59] calculated in their model for an oxygen-blown gasification process a similar hydrogen gas yield of 102 g/kg dry biomass.

Tock and Marechal [60] compared in their thermo-economic modeling the hydrogen production from natural gas and from biomass with regard to energy efficiency optimization by polygeneration of hydrogen, heat and power and including CO_2 capture. They concluded that the system performance is improved by process integration maximizing the heat recovery and valorizing the waste heat. They calculated an energy efficiency of 60% for hydrogen production by biomass gasification compared to 80% efficiency for hydrogen production by steam reforming of natural gas.

Shabani et al. [61] investigated hydrogen production by gasification of rice husk combined with heat recovery for electricity generation by two Rankine cycles. They used a GE gasifier operated at 1200°C and 3 MPa with oxygen as the gasification medium in their model. Hydrogen efficiency of 40% and net electrical efficiency of 3.25% has been calculated if CO_2 is not compressed for capture and storage. With CO_2 capture and storage the net electrical efficiency drops to 1.5%.

Abuadala and Dincer [62] modeled a quite complex integrated system for polygeneration of hydrogen, heat, and power. They considered a system comprising steam gasification of sawdust, a coupled SOFC-SOEC (solid oxide electrolyzer cell), steam reformer, water gas shift reactor, compressors, gas turbine, and burner. They concluded that their results give an indication for hydrogen production costs which are quite favorable and have potential for practical applications.

Beside the aforementioned theoretical studies, an experimental investigation of an efficient integrated process to produce hydrogen is developed in the frame of the UNIfHY research project funded by the fuel cells and hydrogen joint undertaking (FCH-JU) of the European Commission [63]. The system developed in this project comprises a UNIQUE gasifier (described in Section 4.1), a water gas shift reactor, and a pressure swing adsorption unit. Purge gas is recirculated to the gasifier to be used as a heat source for the endothermic reactions in the process. It is predicted that the process with the integration of several subsystems will achieve hydrogen conversion efficiencies higher than 66% [63].

REFERENCES

[1] Obernberger I, Thek G. Basic information regarding decentralized CHP plants based on bio-mass combustion in selected IEA partner countries, IEA Bioenergy—Task 32 Report 2004, <http://www.ieabcc.nl/publications/IEA-CHP-Q1-final.pdf> [accessed 15.10.13].

[2] Ahrenfeldt J, Thomsen TP, Henriksen U, Clausen LR. Biomass gasification cogeneration—a review of state of the art technology and near future perspectives. Appl Therm Eng 2013;50: 1407–17.

[3] Negro SO, Suurs RAA, Hekkert MP. The bumpy road of biomass gasification in the Netherlands: explaining the rise and fall of an emerging innovation system. Technol Forecast Soc Change 2008;75/1:57–77.

[4] Zhou Z, Yin X, Xu J, Ma L. The development situation of biomass gasification power genera-tion in China. Energy Policy 2012;51:52–7.

[5] Kirkels AF, Verbong GPJ. Biomass gasification: still promising? A 30-year global overview. Renewable Sustainable Energy Rev 2011;15:471–81.

[6] Teislev B. Wood chips gasifier combined heat and power, <http://ieatask33.org/app/webroot/files/file/publications/WoodchipsGasifierCombinedheatandPower.pdf> [accessed 14.02.16].

[7] Brandin J, Tuner M, Odenbrand I. Small scale gasification: gas engine CHP for biofuels, Swedish Energy Agency Report 2010, <http://ieatask33.org/app/webroot/files/file/publications/new/Small%20Small_scale_gasification_overview.pdf> [accessed 14.02.16].

[8] Bolhàr-Nordenkampf M, Rauch R, Bosch K, Aichernig C, Hofbauer H. Biomass CHP plant Güssing—using gasification for power generation. In: International conference on biomass utilisation, Thailand, June 2002, <http://www.ficfb.at/pub.htm> [accessed 14.02.16].

[9] Kramreiter R, Url M, Kotik J, Hofbauer H. Experimental investigation of a 125 kW twin-fire fixed bed gasification pilot plant and comparison to the results of a 2 MW combined heat and power plant. Fuel Process 2008;89:90–102.

[10] Rauch R. Personal communication 2014.

[11] <http://www.swu.de/> [accessed 14.02.16].

[12] Minchener AJ. Coal gasification for advanced power generation. Fuel 2005;84:2222–35.

[13] Trevino Coca M. Integrated gasification combined cycle technology: IGCC, <http://www.elcogas.es/en/igcc-technology> [accessed 14.02.16].

[14] Giuffrida A, Romano MC, Lozza G. Efficiency enhancement in IGCC power plants with air-blown gasification and hot gas clean-up. Energy 2013;53:221–9.

[15] Stahl K, Neergaard M. IGCC power plant for biomass utilization, Värnamo, Sweden. Biomass Bioenergy 1998;15:205–11.

[16] Bengtsson S. The CHRISGAS project. Biomass Bioenergy 2011;35:S2–S7.

[17] Arena U, Di Gregorio F, Santonastasi M. A techno-economic comparison between two design configurations for a small scale, biomass-to-energy gasification based system. Chem Eng J 2010;162:580–90.

[18] Aravind PV, de Jong W. Evaluation of high temperature gas cleaning options for biomass gas-ification product gas for Solid Oxide Fuel Cells. Prog Energy Combust Sci 2012;38:737–64.

[19] Bang-Møller C, Rokni M, Elmegaard B, Ahrenfeldt Henriksen UB. Decentralized combined heat and power production by two-stage biomass gasification and solid oxide fuel cells. Energy 2013;58:527–37.

[20] Doherty W, Reynolds A, Kennedy D. Computer simulation of a biomass gasification-solid oxide fuel cell power system using Aspen plus. Energy 2010;35:4545–55.

[21] Nagel FP, Schildhauer TJ, McCaughey N, Biollaz SMA. Biomass-integrated gasification fuel cell systems—Part 2: economic analysis. Int J Hydrogen Energy 2009;34:6826–44.

[22] Colpan CO, Hamdullahpur F, Dincer I. Solid oxide fuel cell and biomass gasification systems for better efficiency and environmental impact. In: Stolten D, Grube T, editors. Proceedings of the 18th world hydrogen energy conference 2010. ISBN: 978-3-89336-651-4. p. 305–13.

[23] Wongchanapai S, Iwai H, Saito M, Yoshida H. Performance evaluation of an integrated small-scale SOFC biomass gasification power generation system. J Power Sources 2012;216:314–22.

[24] Di Carlo A, Borello D, Bocci E. Process simulation of a hybrid SOFC/mGT and enriched air/steam fluidized bed gasifier power plant. Int J Hydrogen Energy 2013;38(14):5857–74.

[25] Di Carlo A, Bocci E, Naso V. Process simulation of a SOFC and double bubbling fluidized bed gasifier power plant. Int J Hydrogen Energy 2013;38(1):532–42.

[26] Bang-Møller C, Rokni M, Elmegaard B. Exergy analysis and optimization of a biomass gasification, solid oxide fuel cell and micro gas turbine hybrid system. Energy 2011;36:4740–52.

[27] Morandin M, Marechal F, Giacomini S. Synthesis and thermo-economic design optimization of wood-gasifier-SOFC systems for small scale applications. Biomass Bioenergy 2013;49:299–314.

[28] Nagel FP, Ghosh S, Pitta C, Schildhauer TJ, Biollaz S. Biomass integrated gasification fuel cell systems. Biomass Bioenergy 2011;35:354–62.

[29] Jewulski J, Stepien M, Blesznowski M, Nanna F. Slip stream testing with a SOFC unit at Güssing and Trisaia plants, <http://www.uniqueproject.eu/public/deliverables/D7.2.pdf> [accessed 14.02.16].

[30] Martini S, Kleinhappl M, Hofbauer H. Gasaufbereitung und SOFC-Versuchsbetrieb mit produktgas einer Wirbelschicht-Dampf-Vergasung, DGMK Tagungsbericht 2010. ISBN: 978-3-941721-06-7.

[31] Hofmann P, Schweiger A, Fryda L, Panopoulos KD, Hohenwarter U, Bentzen JD, et al. High temperature electrolyte supported Ni-GDC/YSZ/LSM SOFC operation on two-stage Viking gasifier product gas. J Power Sources 2007;173:357–66.

[32] Hofmann P, Panopoulos KD, Fryda LE, Schweiger A, Ouweltjes JP, Karl J. Integrating biomass gasification with solid oxide fuel cells: effect of real product gas tars, fluctuations and particulates on Ni-GDC anode. Int J Hydrogen Energy 2008;33:2834–44.

[33] Biollaz S, Hottinger P, Pitta C, Karl J. Results from a 1200 hour test of a tubular SOFC with wood gas. In: Proceedings of 17th European biomass conference and exhibition 2009, Hamburg, Germany.

[34] Kopyscinski J, Schildhauer TJ, Biollaz SMA. Production of synthetic natural gas (SNG) from coal and dry biomass—a technology review from 1950 to 2009. Fuel 2010;89:1763–83.

[35] Gröbl T, Walter H, Haider M. Biomass steam gasification for production of SNG—process design and sensitivity analysis. Appl Energy 2012;97:451–61.

[36] Van der Meijden CM, Veringa HJ, Rabou LPLM. The production of synthetic natural gas (SNG): a comparison of three wood gasification systems for energy balance and overall efficiency. Biomass Bioenergy 2010;34:302–11.

[37] Biollaz S, Schildhauer TJ, Ulrich D, Tremmel H, Rauch R, Koch M. Status report of the demonstration of BioSNG production on a 1 MW SNG scale in Güssing. In: Proceedings of 17th European biomass conference and exhibition 2009, Hamburg, Germany.

[38] Vitasari CR, Jurascik M, Ptasinski KJ. Exergy analysis of biomass-to-synthetic natural gas (SNG) process via indirect gasification of various biomass feedstock. Energy 2011;36:3825–37.

[39] Wirth S, Markard J. Context matters: how existing sectors and competing technologies affect the prospects of the Swiss Bio-SNG innovation system. Technol Forecast Soc Change 2011;78:635–49.

[40] Gunnarsson I. Efficient transfer of biomass to Bio-SNG of high quality: the GoBiGas project, Nordic Baltic Bioenergy 2013, Oslo, Norway, <http://nobio.no/upload_dir/pics/Ingemar-Gunnarsson.pdf> [accessed 14.02.16].

[41] <http://gobigas.goteborgenergi.se> [accessed 14.02.16].

[42] European Biofuels Technology Platform, EBTP Newsletter 23, <http://biofuelstp.eu/bio-sng.html>; December 2015 [accessed 14.02.16].

[43] Fredriksson Moller B, Molin A, Stahl K. Bio2G—a full-scale reference plant in Sweden for production of bio-SNG (biomethane) based on thermal gasification of biomass. In: TC biomass conference 2013, Chicago, USA, <http://www.gastechnology.org/tcbiomass/Pages/2013-presentations.aspx> [accessed 14.02.16].

[44] Shuying L, Guocai W, DeLaquil P. Biomass gasification for combined heat and power in Jilin province, People's Republic of China. Energy Sustainable Dev 2001;V(1):47–53.

[45] Steubing B, Zah R, Ludwig C. Life cycle assessment of SNG from wood for heating, electricity, and transportation. Biomass Bioenergy 2011;35:2950–60.

[46] Felder R, Dones R. Evaluation of ecological impacts of synthetic natural gas from wood used in current heating and car systems. Biomass Bioenergy 2007;31:403–15.

[47] Fahlen E, Ahlgren EO. Assessment of integration of different biomass gasification alternatives in a district-heating system. Energy 2009;34:2184–95.

[48] Ahman M. Biomethane in the transport sector—an appraisal of the forgotten option. Energy Policy 2010;38:208–17.

[49] Gassner M, Marechal F. Thermo-economic process model for thermochemical production of Synthetic Natural Gas (SNG) from lignocellulosic biomass. Biomass Bioenergy 2009;33:1587–604.

[50] Difs K, Wetterlund E, Trygg L, Söderström M. Biomass gasification opportunities in a district heating system. Biomass Bioenergy 2010;34:637–51.

[51] Narvaez A, Chadwick D, Kershenbaum L. Small-medium scale polygeneration systems: methanol and power production. Appl Energy 2014;113:1109–17.

[52] Haro P, Ollero P, Villanueva Perales AL, Gomez-Barea A. Thermochemical biorefinery based on dimethyl ether as intermediate: technoeconomic assessment. Appl Energy 2013;102:950–61.

[53] Meerman JC, Ramirez A, Turkenburg WC, Faaij APC. Performance of simulated flexible integrated gasification polygeneration facilities, part B: economic evaluation. Renewable Sustainable Energy Rev 2012;16:6083–102.

[54] Djuric Ilic D, Dotzauer E, Trygg L, Broman G. Introduction of large-scale biofuel production in a district heating system—an opportunity for reduction of global greenhouse gas emissions. J Cleaner Prod 2014;64:552–61.

[55] Wetterlund E, Leduc S, Dotzauer E, Kindermann G. Optimal localisation of biofuel production on a European scale. Energy 2012;41:462–72.

[56] Haro P, Trippe F, Stahl R, Henrich E. Bio-syngas to gasoline and olefins via DME—a comprehensive techno-economic assessment. Appl Energy 2013;108:54–65.

[57] Ng KS, Sadhukhan J. Techno-economic performance analysis of bio-oil based Fischer-Tropsch and CHP synthesis platform. Biomass Bioenergy 2011;35:3218–34.

[58] Toonssen R, Woudstra N, Verkooijen AHM. Exergy analysis of hydrogen production plants based on biomass gasification. Int J Hydrogen Energy 2008;33:4074–82.

[59] Bhattacharya A, Bhattacharya A, Datta A. Modeling of hydrogen production process from biomass using oxygen blown gasification. Int J Hydrogen Energy 2012;37:18782–90.

[60] Tock L, Marechal F. H_2 processes with CO_2 mitigation: thermo-economic modeling and process integration. Int J Hydrogen Energy 2012;37:11785–95.

[61] Shabani S, Delavar MA, Azmi M. Investigation of biomass gasification hydrogen and electricity co-production with carbon dioxide capture and storage. Int J Hydrogen Energy 2013;38:3630–9.

[62] Abuadala A, Dincer J. Exergoeconomic analysis of a hybrid system based on steam biomass gasification products for hydrogen production. Int J Hydrogen Energy 2011;36:12780–93.

[63] <http://www.unifhy.eu> [accessed 14.02.16].

Chapter 8

Conclusions and Outlook

Biomass is one of the major renewable energy sources beside wind and solar energy. In contrast to wind and solar energy, renewable energy production from biomass can be adjusted to current consumption need. Gasification is considered as a key technology for the use of biomass offering high flexibility and efficiency. To promote this technology in the future, advanced, cost-effective, and highly efficient gasification processes and systems are required, including gas cleaning and conditioning of the raw produced gas for subsequent utilization.

New knowledge and efficient and cost-competitive industrial applications are required for the long run, showing evidence of promising developments and cross-fertilization with sectors other than energy, which may provide ideas, experiences, technology contributions, new approaches, innovative materials, and skills.

This book has described some very interesting new concepts and strategies in biomass gasification. New advanced process integration and combination concepts enable higher process efficiencies, better gas quality and purity, and lower investment costs. Some of these concepts might provide the opportunity to overcome obstacles which have inhibited the achievement of a higher market share for biomass gasification so far.

The recently developed UNIQUE gasifier concept which integrates gasification, gas cleaning, and conditioning in one reactor unit is a nice example of an innovative new integration concept. This concept offers a compact and simple design, decreases process complexity, and requires less footprint as well as reduced capital expenditures. The overall efficiency of the conversion process is high, as no cooling and reheating for gas cleaning is required. Options for further process integrations in the UNIQUE gasifier concept are given by the additional integration of gas separation membranes in the system. Additional integration of ionic oxygen transport membranes into the fluidized bed of the gasifier could be a way for an efficient and economic supply of oxygen as a gasifying agent into the gasifier. Moreover, hydrogen separation membranes could be integrated into the freeboard of the gasifier for selective separation of hydrogen from the produced gas.

New advanced multistage gasification concepts separate and combine pyrolysis and gasification in single controlled stages. In this way pyrolysis and

Advanced Biomass Gasification. DOI: http://dx.doi.org/10.1016/B978-0-12-804296-0.00008-7

gasification can be improved by optimizing the operating conditions in each stage. This enables the achievement of high process efficiencies and a syngas with low tar concentration. On the other hand, the complexity of the process is increased by combining different reactors.

A special approach performs pyrolysis and gasification at different locations. The concept aims to produce concentrated oil–char slurries by decentralized pyrolysis plants and gasification of the slurries and production of biofuels in a large centralized plant. In this way, transportation of the biomass as well as biofuel production becomes more economical.

Other concepts combine gasification with a combustion stage. In this way unreacted char can be combusted to increase the overall process efficiency. Additionally, by combining gasification with a partial oxidation stage, tar can be converted. A product gas with a high heating value can be generated by steam gasification using the dual fluidized bed process with internal combustion.

An easy and cost-effective way to reduce fossil CO_2 emissions is the indirect cofiring of biomass in coal-fired boilers by gasification of the biomass. This concept is well proven by the operating experience of biomass gasifiers with capacities up to $140\,MW_{th}$ in large coal-fired power plants.

New gasification concepts, such as supercritical water gasification, provide interesting advantages for special kinds of biomass. Oxygen–steam gasification for producing syngas with high heating value, which is state-of-the-art in coal gasification, is still under investigation for biomass gasification. Chemical looping using oxygen carrier material as well as the use of ionic oxygen transport membranes are interesting concepts for supplying oxygen for the gasification of the biomass.

Polygeneration strategies for the production of more than one product in a combined process can significantly improve process efficiency, economic viability, and sustainability of the use of biomass. Combined heat and power production is a classic example for a polygeneration process, but also new approaches, such as combined synthetic natural gas, heat and power production, or biofuels, heat and power production, as well as hydrogen and heat production attract increasing interest. Furthermore, these facilities offer the flexibility to produce electricity when it is needed and if it is not needed to produce biofuels or chemicals from the syngas. This flexibility is not free of charge. It is coupled to higher investments in both the reactors as well as in the power generation equipment. Back-up power plants are needed and the required investments can be made in polygeneration plants. However, development and realization of polygeneration strategies is highly influenced by national governmental energy strategies.

The development of innovative catalysts, sorbents, and high temperature filtration media represent a fundamental requirement to increase yield and purity of the produced biomass gasification gas, to allow efficient conversion into power (high temperature fuel cells, gas turbines, advanced, strictly integrated heat and power plant installations), as well as further catalytic processing addressed to the production of second-generation biofuels (liquid fuels for transportation, hydrogen) and chemicals.

High temperature gas cleaning and catalytic conditioning is the focal point to promote industrial applications of biomass for energy and chemicals. To avoid an unacceptable increase of plant and operating costs, gas cleaning and conditioning has to be integrated with biomass conversion and carried out at the same close temperature range, to preserve the thermal energy content of the biomass gas.

Another important aspect for future developments, which is not discussed in very much detail in this book, is the need to enlarge the biomass feedstock base for advanced gasification processes. The use of biogenic residuals from agriculture and forestry would avoid conflicts with the food supply and other land uses. Nevertheless, competitive energetic and nonenergetic utilization of these resources still has to be addressed. The type of solid feedstock has a significant impact on the technology for gasification. Since especially biomass residues have very often rather low ash fusion temperatures and release high amounts of detrimental trace elements during gasification, such behavior has to be considered when it comes to the development of new and advanced technological solutions.

Index

Printed in the United States
By Bookmasters